A Short Treatise on Shock Waves and Equations of State

A Short Treatise on Shock Waves and Equations of State

Y. K. Huang, Ph. D.
Formerly U. S. Ballistic Research Laboratory
Aberdeen Proving Ground, Maryland

Nova Science Publishers, Inc.
New York

Senior Editors: Susan Boriotti and Donna Dennis
Coordinating Editors: Tatiana Shohov and Jennifer Vogt
Office Manager: Annette Hellinger
Graphics: Wanda Serrano
Editorial Production:, Matthew Kozlowski, Jonathan Rose and Maya Columbus
Circulation: Ave Maria Gonzalez, Indah Becker and Vladimir Klestov
Communications and Acquisitions: Serge P. Shohov
Marketing: Cathy DeGregory

Library of Congress Cataloging-in-Publication Data
Available upon request.
ISBN 1-59033-324-1

400 Oser Ave, Suite 1600
Hauppauge, New York 11788-3619
Tele. 631-231-7269 Fax 631-231-8175
e-mail: Novascience@earthlink.net
Web Site: http://www.novapubishers.com

Printed in the United States of America

Contents

Preface

Chapter 1 Introduction 1

1.1 Origin of shock waves (SW) 1

1.2 A catalog of SW 2

1.3 Generation of SW in laboratory 3

1.4 Basic shock jump conditions 4

1.5 Equations of state (EOS) 6

1.6 A thermodynamic interplay between EOS and SW 7

1.7 Shock behavior and properties with an arbitrary EOS 10

Chapter 2 Theory of SW in γ-law gases 13

2.1 Normal shocks 13

2.2 Oblique shocks 16

2.3 Spherical and cylindrical blast waves 22

2.4 Normal shock reflection and transmission 26

2.5 Oblique shock reflection and diffraction 32

2.6 Summary 39

Chapter 3 Theoretical and empirical EOS 41

3.1 The four states of matter 41

3.2 A theory for the construction of EOS 42

3.3 Classical ideal gases 45

3.4 Quantum ideal gases 49

3.5 Van der Waals fluids 53

3.6 Tait EOS for liquids 56

3.7 Grüneisen EOS for solids 59

3.8 EOS for plastics, mixtures, powders, and porous material 67

3.9 Thomas-Fermi EOS for dense matter at ultra-high pressure 74

3.10 Summary 79

Chapter 4 Some interdisciplinary SW and EOS 81

4.1 Hydraulic jumps and underwater shocks 81

4.2 Negative shocks and SW involving phase transition 85

4.3 Detonation waves 88
4.4 Ionizing shocks 94
4.5 Radiating shocks 96
4.6 Hydromagnetic shocks 99
4.7 Laser shocks 104

Chapter 5 Special topics 107
5.1 Shock tubes 107
5.2 Hypervelocity impact 111
5.3 Shaped-charge jet penetration 116
5.4 Inertial confinement fusion (ICF) by imploding SW 118
5.5 Cosmic and isothermal shocks 121
5.6 Shock focusing and extracorporeal shock wave lithotripsy (ESWL) 125
5.7 SW in traffic jams 127
5.8 Shock structure and thickness 132
5.9 Computer simulation of shock flow 135
Epilogue 141
References 143
Appendix 147
Index

Preface

Analytically, shock waves (SW) are a result from the study of nonlinear waves in continuua. When large-amplitude waves propagate in nonviscous compressible supersonic flows, the forerunner waves are overtaken by the accelerating follower waves (because of compressibility and nonlinearity) and they eventually merge into a steep wave front known as SW. Literally speaking, a SW is an intense disturbance in an environment of imbalance (e.g. a thunderclap, earthquake, explosion). Today the phenomenology of SW has considerable significance in many aspects of nature and humankind, and there exists a hidden link of SW to equations of state (EOS) which characterize the shock environments. A SW problem ought to have an EOS for its solution, and a new EOS can be derived from unexplored SW data analytically or empirically. Thus it is of relevant interest to study SW and EOS collaterally. This monograph has grown out of many years of study, teaching, and research with such a perspective. If the phrase "science and technology" implies "theory and applications", then this is just such a short handbook in my mind.

From the root to the state of the art, this book is organized in five chapters which contain thirty nine sections, and it should prove to be all essential for an expedient grasp of the fundamentals of shock behavior and properties. A wide spectrum of topics is covered in this short handbook, including most of the major EOS, a theory of SW in seven particular environments, and nine specialized subjects which may fill a good many volumes. The thirty references are not only the author's sources of check for accuracy and update but also a list of quality for the reader's further reading. Included in the appendix are eight reprints of the author's earlier publication, which may be of relevant interest to the study of Chapter 3. An extensive index is prepared for purposeful cross-reference or review. It is hoped that the index need not be superficial. Finally the acronyms, symbols, notations, subscripts, and superscripts are all designated in the text. Yet some symbols may have been used repetitively with different meaning in different context (e.g. B for constant, bulk modulus, or magnetic induction; h for enthalpy or Planck constant $\hbar$). These are unavoidable due to the limited number of alphabets and digits. Presumably the reader is aware of all these and the unit and dimension of a symbol.

This monograph differs from many similar titles in style and difficulty. Our presentation is of brevity and at an intermediate level of difficulty, without invoking the more advanced theory. Accordingly, this book is meant to be useful for studying, teaching, research, and reference. It is a pleasure to have been collaborating with the staff of Nova Science Publishers, Inc., particularly Nadya S. Columbus, the Vice-President, who initiated the Agreement of publishing this book; Jennifer Vogt, Editor, who has done the superb preparation for the press; and Frank Columbus, the President and Editor-in-Chief, whose supervision, consultation, and advices are most helpful and pleasant. Also I am indebted to my wife, Elizabeth, whose graceful forbearance lends me so many hours in absence from her companionship to complete the manuscript of this book. My daughter, Mary, typed the first two chapters of the book, and Eric, my son, spent hundreds of his professional hours for the polished finale. I am greatly appreciative of his neat workmanship!

Y.K. Huang

Latham, New York
Spring 2002

Acknowledgment
Many thanks are due to American Institute of Physics for permissions to reproduce items No. 1, 2, 3, 5, and 7 of Appendix from the respective journals. Special permission is granted by Institute of Physics Publishing Ltd., UK for the reproduction of No. 4 in Appendix (with my indebtedness, the journal's homepage at Internet www.iop.org/journals/jphysd). Also I am grateful to Plenum Publishing Corp. and the Franklin Institute for their permissions to reproduce No. 6 and No. 8 of Appendix, respectively.

The following four figures are reproduced with permission by or credit to the respective sources as listed below:

Fig. 5.7 on p. 126 of this book, from R.A. Riehle, Jr. and R.C. Newman (ed.): Principles of Extracorporeal Shock Wave Lithotripsy (Churchill Livingstone, 1987), p. 14.

Fig. 5.8 on p. 126, with permission granted by Kirk A. Wilks, Director International

Marketing Communications, Dornier Medical Systems, Inc., Kennesaw, GA 30144.

Fig. 5.13 on p. 139, courtesy American Institute of Physics (from AIP Conference Proceedings No. 208 The 17th Int. Symp. Shock Waves and Shock Tubes (1989), p. 251).
Fig. 5.14 on p. 139, courtesy Los Alamos National Laboratory (from the laboratory report LA-3466 (1966), pp. 121, 122).

Chapter 1 Introduction

1.1 Origin of Shock Waves

What is a shock wave (SW)? How does it happen? Why? Where, when and for what? Indeed it can serve as a good start of our inquiry to answer these questions. SW originate from the high-speed, inviscid fluid flows in many fields of study such as acoustics, gasdynamics, astrophysics, etc. Remarkably, the pertinent large-amplitude waves tend to distort into a coalition known as SW. As the forerunner waves are overtaken by the accelerating follower waves (because of compressibility and nonlinearity), they eventually merge into a steep front. This is the description of shock formation by the classical piston problem, although the structure of SW is a rather different story which involves viscosity and diffusivity. Literally speaking, a SW is an intense disturbance in an environment of imbalance. It may occur to shock us in awe (e.g. thunderclap, earthquake, explosion); it propagates at supersonic speed; and many SW happen to be so intense as of mega (10^6) or tera (10^{12}) magnitudes, irreversibly beyond control. That is the nature of SW.

Shock phenomena are ubiquitous; they take place in all states of matter (gas, liquid, solid, and plasma), and they happen in all time (past, present, and future). In the very beginning, big bangs create galaxies, planets, and the earth. Novae and supernovae are historical events of SW. Today technology and medicine are making use of SW to promote our well-being considerably. Thus artificial diamonds have been made by shock compaction of graphite powder; kidney stones are pulverized by focusing underwater SW without involving the risks of surgery; and an abundant energy supply will eventually be available by the inertial confinement fusion (ICF) of deuterium and tritium. Soon the 21st century will unfold more "future shocks" of this world of science and technology. For better or worse, SW have far-reaching consequences globally. It is no exaggeration that the sharp edge of SW can cause terrible hazards and disasters. May SW be the genie of all world peace!

1.2 A Catalog of Shock Waves

There exists a world of SW in various environments and specialized fields of study. Let us enlist a catalog of SW via forty examples below.

A. Shock phenomena in air: (1) hand clapping, balloon collapse; (2) cowboy's bull whip; (3) thunderclap, lightning, arc-discharge; (4) blast waves from bombing or explosion; (5) ballistic shock of a flying bullet; (6) sonic boom of a supersonic aircraft; (7) bow shock of a spacecraft's reentry.

B. Shock waves in other gases: (8) steam-turbine nozzle flow; (9) burnt gases from gun muzzle, rocket, jet engine; (10) argon, hydrogen or helium in shock tube; (11) ICF of deuterium and tritium by laser imploding shock.

C. Shock waves in water: (12) hydraulic jump; (13) waterhammer; (14) surf, breaker, bore; (15) flash flood; (16) depth-charge explosion; (17) micro-implosion of hydrodynamic cavitation around a ship's propeller; (18) water-entry of a missile, landing impact of a seaplane; (19) submerged electric-arc discharge to pulverize kidney stones.

D. Shock waves in solid: (20) earthquake, aftershock, underground nuclear testing; (21) ballistic impact, hypervelocity impact; (22) SW induced by contact explosion or laser irradiation; (23) shaped-charge penetration of armor, anti-tank warhead, anti-aircraft missile; (24) shock compression of metal, porous material, composite for EOS study; (25) electromagnetic SW in nonlinear ferromagnetic circuits.

E. Shock waves in plasma: (26) magnetohydrodynamic (MHD) shocks in cold plasma; (27) ionizing SW according to Saha's theory; (28) SW in hot plasma ($T>5000^oK$); (29) collisionless SW in rarefied gas or outer space; (30) exploding wire and foil; (31) plasma shock trailing the moon.

F. Spectacular shock phenomena: (32) bow shock due to solar wind interacting with the earth's magnetosphere; (33) solar flare, sunspot activities, geomagnetic storm; (34) giant meteoric impact on moon's or earth's surface; (35) stellar collision, supernova; (36) galaxy explosion with energy yield equivalent to 10^{49} tons of TNT; (37) relativistic shocks in ultradense matter, nuclear and cosmic events.

G. Shock waves for industry: (38) explosive working of metal, spacecraft forming, explosive welding, powder compaction; (39) laser cutting and drilling; and

(40) chemical synthesis using catalytic shocks to convert refuse into fuel.

Strictly speaking, SW are not easily classifiable. Thus many of the above-mentioned examples may belong to: (a) aerodynamic, gasdynamic, or hydrodynamic shocks; (b) inert, explosive, or reactive shocks; (c) electric, gravity, kinematic shocks; (d) plane, curved, spherical, or cylindrical shocks of the diverging or converging type; and (e) positive or compression shocks. Other categories are not so well known, namely, negative or rarefaction shocks in retrograde fluids of van der Waals EOS, imploding shocks for ICF, laser-supported detonation (LSD) waves, and isothermal SW in stellar explosion. All these SW are to be characterized by three jump conditions including the Rankine-Hugoniot energy conservation with irreversible entropy rise.

Also, it is worth mentioning that traffic congestion, explosive publication, shocking news, and other similar events are SW by analogy. They propagate like SW but lack any link to the Rankine-Hugoniot relation. From a specialized point of view, some SW pose difficult problems to solve; however, we will treat most of the fundamental problems at an intermediate level of difficulty, without invoking the more advanced study. This is our criterion to define the scope of this book.

1.3 Generation of SW in laboratory

From the foregoing catalog we see that SW occur in various environments from time to time, often lacking control. For research and development SW must be produced or reproduced in laboratory, with measurable properties and effects for assessment. Conventional shock tubes are a versatile research tool for studying gasdynamic SW (see also Section 5.1) in science and technology. Yet the latest trend is to generate strong SW as exemplified by: (a) $M = 7$ in helium, (b) $M = 10$ in hydrogen, (c) $M = 20$ in combustion gas, (d) $M > 20$ in multi-stage shock tubes, (e) $M > 100$ by electromagnetic drive, and (f) $M \geq 1000$ in ionized deuterium for thermonuclear research. Note $M = U/c$ = Mach number (U = shock velocity and c = sound velocity). The larger the Mach number, the stronger the shock at high velocity and/or high temperature ($M > 1$ supersonic and $M > 5$ hypersonic).

During the past 50 years or so, intensive research of solids under shock compression has been conducted for the study of their EOS and high-pressure properties. Again the new trend is to produce SW in solids by: (a) contact explosion $p \sim 0.5$ Mb, (b) flyer-plate impact from high explosive $p \sim 1.0$ Mb, (c) projectile impact from light gas gun $p \sim 10$ Mb, (d) plate impact from electric gun $p \sim 50$ Mb, (e) plate impact from rail gun $p \sim 100$ Mb, (f) LSD or laser irradiation $p \sim 100$ Mb also at high temperature, (g) underground nuclear test $p \sim 1000$ Mb (high temperature 10^6K, too). Note 1 Mb = 10^6 bars or atmospheres. Thomas-Fermi EOS applies to ultra-high pressures from 100 to 10^5 Mb.

Liquids are either gas-like or solid-like, and their shock properties have been studied with samples enclosed in containers subjected to contact explosion or plate impact. Plasma shocks are studied in shock tubes with magnetic field affecting the inert SW as magnetohydrodynamic shocks (plasma as a quasi-neutral, conducting fluid). For partially and fully-ionized gases, kinetic theory is needed for the investigation of SW in hot plasma.

1.4 Basic shock jump conditions

The original formulation of a SW is a discontinuity in a one-dimensional supersonic flow, across which the upstream state (with subscript 1) jumps to the downstream state (with subscript 2) according to the conservation of mass, momentum, and energy. These jump conditions are expressed as three simple equations with reference to a coordinate frame moving with the shock front. Thus the shock front is treated as stationary and the fluid flow is reversed with relative velocity $u_i = U - \mu_i$ (U = shock velocity, μ = mass flow velocity, $i = 1,2$). See Fig. 1.1. Let p, ρ, E, and h denote the pressure, density, specific internal energy, and enthalpy respectively. Accordingly, the shock jump conditions may be expressed as

$$\rho_1 u_1 = \rho_2 u_2 = m \tag{1.1}$$

$$p_1 + \rho_1 u_1^2 = p_2 + \rho_2 u_2^2 \tag{1.2}$$

$$h_1 + \tfrac{1}{2}u_1^2 = h_2 + \tfrac{1}{2}u_2^2 \tag{1.3}$$

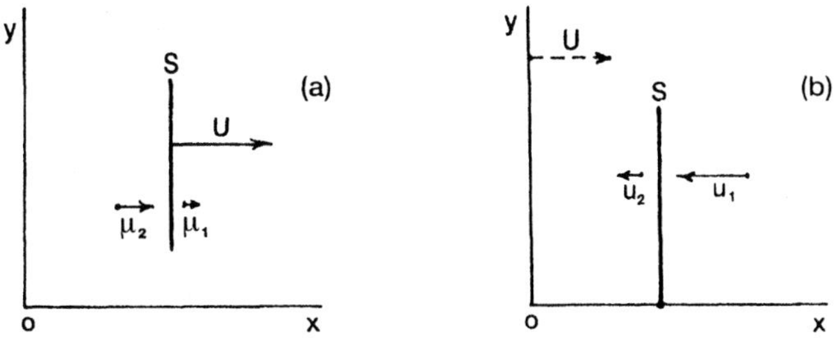

Fig. 1.1 Fluid flow with reference to (a) laboratory coordinates, (b) moving coordinates.

Here the SW is understood to be inert, without involving chemical reaction, phase transition, magnetic field, or radiation. With $h = E + pv$ and $v = 1/\rho$, we may readily combine the above three equations to give

$$p_2 - p_1 = m(u_1 - u_2) = m^2(v_1 - v_2)$$

and

$$h_2 - h_1 = \tfrac{1}{2}(u_1 - u_2)\cdot(u_1 + u_2) = \frac{1}{2m}(p_2 - p_1)\cdot m(v_1 + v_2).$$

Thus from Eqs. (1.1) - (1.3), we obtain

$$u_1^2 = (mv_1)^2 = -v_1^2\left(\frac{p_2 - p_1}{v_2 - v_1}\right) \tag{1.4}$$

$$U = v_1\sqrt{\frac{p_2 - p_1}{v_1 - v_2}}, \text{ with } \mu_1 = 0 \tag{1.4a}$$

$$(u_1 - u_2)^2 = (p_2 - p_1)(v_1 - v_2) \tag{1.5}$$

$$h_2 - h_1 = \tfrac{1}{2}(v_1 + v_2)(p_2 - p_1) \tag{1.6}$$

$$E_2 - E_1 = \tfrac{1}{2}(p_1 + p_2)(v_1 - v_2) \tag{1.7}$$

Eqs. (1.6) and (1.7) are known as the Rankine-Hugoniot relation for SW (normal or oblique), and they have the thermodynamic counterparts $dh = vdp$ and $dE = -pdv$, respectively, for a continuous, reversible, isentropic change (entropy $s_1 = s_2$ vs. $s_2 > s_1$

for SW). Moreover, Eqs. (1.4) - (1.5) correspond to

$$c^2 = -v^2 \frac{dp}{dv} = \left(\frac{\partial p}{\partial \rho}\right)_s \text{ , and } du = \sqrt{-dpdv} = c\frac{d\rho}{\rho} = \frac{dp}{\rho c}$$

for the sound velocity and the acoustic impedance $\left(\rho c = dp/du = \sqrt{-dp/dv}\right)$, respectively.

So far we have only introduced the basic equations of a normal shock jump, although the other missing link is an EOS to be discussed subsequently. It will become clear in Chapter 2 and 4 that these jump conditions can be generalized or used by analogy to treat other forms of SW.

1.5 Equations of state

Thermodynamically speaking, an EOS is the correlation between state variables such as $p = p(v,\mathrm{T})$, T being the temperature. Its rigorous derivation may be sought from statistical mechanics as $p = -(\partial \mathrm{F}/\partial v)_\mathrm{T} = k\mathrm{T}(\partial ln\mathrm{Z}/\partial \mathrm{v})_\mathrm{T}$ (see Eq. (3.8) for detail). However, we can take a simpler approach by the method of thermodynamics as follows.

Let us introduce the basic equations of thermodynamics:

$$\mathrm{T}ds = d\mathrm{E} + pdv = dh - vdp \tag{1.8}$$

$$\left(\frac{\partial \mathrm{E}}{\partial v}\right)_\mathrm{T} = \mathrm{T}\left(\frac{\partial p}{\partial \mathrm{T}}\right)_v - p \text{ , } \left(\frac{\partial \mathrm{E}}{\partial \mathrm{T}}\right)_v = \mathrm{T}\left(\frac{\partial s}{\partial \mathrm{T}}\right)_v = c_v \tag{1.8a}$$

$$\left(\frac{\partial h}{\partial p}\right)_\mathrm{T} = v - \mathrm{T}\left(\frac{\partial v}{\partial \mathrm{T}}\right)_p \text{ , } \left(\frac{\partial h}{\partial \mathrm{T}}\right)_p = \mathrm{T}\left(\frac{\partial s}{\partial \mathrm{T}}\right)_p = c_p \tag{1.8b}$$

where s denotes the entropy, c_v the isometric specific heat, and c_p the isopiestic specific heat. Given $p = p(v,\mathrm{T})$, we can seek

$$\mathrm{E}(v,\mathrm{T}) = \int^\mathrm{T} c_v d\mathrm{T} + \int^v \left[\mathrm{T}\left(\frac{\partial p}{\partial \mathrm{T}}\right)_v - p\right]dv \tag{1.9}$$

Eliminating T between $p(v,\mathrm{T})$ and $\mathrm{E}(v,\mathrm{T})$ gives $\mathrm{E} = \mathrm{E}(p,v)$. Similarly, given v =

$v(p,\mathrm{T})$, we can deduce

$$h(p,\mathrm{T}) = \int^{\mathrm{T}} c_p d\mathrm{T} + \int^{p} \left[v - \mathrm{T}\left(\frac{\partial v}{\partial \mathrm{T}}\right)_p \right] dp \tag{1.10}$$

and hence $h = h(p,v)$. For illustration let us consider the ideal gas EOS $pv = \mathrm{RT}$, with R denoting the gas constant. Assuming c_v and c_p to be constant in Eqs. (1.9) and (1.10), we obtain

$$c_p\mathrm{T} = h = \mathrm{E} + pv = (c_v + \mathrm{R})\mathrm{T}$$

$$c_v = \frac{\mathrm{R}}{\gamma - 1} \text{ with } \gamma = c_p / c_v = \text{constant}$$

$$\mathrm{E} = c_v\mathrm{T} = \frac{\mathrm{RT}}{\gamma - 1} = \frac{pv}{\gamma - 1} \tag{1.11}$$

$$h = c_p\mathrm{T} = \frac{\gamma\mathrm{RT}}{\gamma - 1} = \frac{\gamma pv}{\gamma - 1} \tag{1.12}$$

Substituting Eq. (1.11) in Eq. (1.8) and letting $ds = 0$, we immediately obtain the isentropic EOS

$$p_s v^{\gamma} = \text{const.} \quad \mathrm{T}_s v^{\gamma - 1} = \text{const.} \tag{1.13}$$

The ideal gas is elsewhere referred to as the γ-law gas in view of Eqs. (1.11) and (1.13), which can categorically represent other gases with constant $\gamma \neq c_p/c_v$. The general definition of γ is given by Eq. (3.1). See also Eq. (3.13) in general. Chapters 2 and 3 contain more details about γ modified.

1.6 A thermodynamic interplay between EOS and SW

The basic shock jump conditions are previously represented by Eqs. (1.1) - (1.3), but they contain four unknowns (p, ρ, u, and h). Accordingly a fourth equation is needed for the simultaneous solution of the downstream variables in terms of the upstream quantities. Let this be supplied by a thermodynamic EOS, $h = h(p,v)$ with $v = 1/\rho$. Then we can deduce a shock EOS, $p_H = p_H(v)$, by combing $h = h(p,v)$ and Eq. (1.6), and hence Eqs. (1.4) and (1.5) are suitable for the solution of u_2 and v_2 in terms of p_1, v_1, and u_1. More details are given in Chapter 2. Conversely, given a shock EOS, we can substitute it into Eq. (1.8) to deduce other relations, say $p_S = p_S(v)$ with $ds = 0$. Such an interplay between SW and EOS may be illustrated below.

It is very interesting now to re-write Eq. (1.7) as

$$\mathrm{E} - \mathrm{E}_o = \tfrac{1}{2}(p_H + p_o)(v_o - v)$$

with the upstream state (p_o, v_o, T_o, E_o all being constant) and the downstream state (p_H, v, T, E) all varying. Thus we may substitute

$$d\mathrm{E} = \tfrac{1}{2}(v_o - v)dp_H - \tfrac{1}{2}(p_H + p_o)dv$$

in Eq. (1.8) to obtain

$$2\mathrm{T}\frac{ds}{dv} = (v_o - v)\frac{dp_H}{dv} + (p_H - p_o)$$

$$2\left(\mathrm{T}\frac{d^2s}{dv^2} + \frac{d\mathrm{T}}{dv}\frac{ds}{dv}\right) = (v_o - v)\frac{d^2p_H}{dv^2}$$

$$2\left(\mathrm{T}\frac{d^3s}{dv^3} + 2\frac{d\mathrm{T}}{dv}\frac{d^2s}{dv^2} + \frac{d^2\mathrm{T}}{dv^2}\frac{ds}{dv}\right) = (v_o - v)\frac{d^3p_H}{dv^3} - \frac{d^2p_H}{dv^2}$$

From these we deduce

$$\left(\frac{ds}{dv}\right)_o = \left(\frac{d^2s}{dv^2}\right)_o = 0 \quad \text{and} \quad \left(\frac{d^3s}{dv^2}\right)_o = -\frac{1}{2\mathrm{T}_o}\left(\frac{d^2p_H}{dv^2}\right)_o$$

For an arbitrary shock EOS $p_H = p_H(v)$, the entropy is an implicit function $s = s(p_H, v) = s(v)$ which may be sought by Taylor series expansion:

$$s = s_o + \left(\frac{ds}{dv}\right)_o (v - v_o) + \frac{1}{2!}\left(\frac{d^2s}{dv^2}\right)_o (v - v_o)^2 + \frac{1}{3!}\left(\frac{d^3s}{dv^3}\right)_o (v - v_o)^3 + \ldots$$

Accordingly the entropy rise across a SW is given by

$$s - s_o = \frac{1}{12T_o}\left(\frac{d^2 p_H}{dv^2}\right)_o (v_o - v)^3 > 0 \tag{1.14a}$$

Similar manipulation with Eqs. (1.6) and (1.8) leads to

$$s - s_o = \frac{1}{12T_o}\left(\frac{d^2 v}{dp_H^2}\right)_o (p_H - p_o)^3 > 0 \tag{1.14b}$$

which is accurate to the third order.

Now that a logical inversion of $s = s(p_H, v)$ permits the formalism $p_H = p_H(v, s)$, we may expand the latter as

$$p_H = p_o + \left(\frac{\partial p}{\partial v}\right)_{s_o}(v - v_o) + \frac{1}{2!}\left(\frac{\partial^2 p}{\partial v^2}\right)_{s_o}(v - v_o) + \frac{1}{3!}\left(\frac{\partial^3 p}{\partial v^3}\right)_{s_o}(v - v_o)^3 + \ldots$$

$$+ \left(\frac{\partial p}{\partial s}\right)_{v_o}(s - s_o) + \ldots$$

$$= p_s + \frac{1}{12T_o}\left(\frac{d^2 p_H}{dv^2}\right)_o \left(\frac{\partial p}{\partial s}\right)_{v_o}(v_o - v)^3 + \ldots$$

to the third degree of accuracey only. Introducing the Grüneisen parameter $\Gamma = v(\partial p/\partial E)_v = (v/T)(\partial p/\partial s)_v$, we deduce

$$p_H = p_s + \frac{\Gamma_o}{12v_o}\left(\frac{d^2 p_H}{dv^2}\right)_o (v_o - v)^3 \tag{1.15}$$

$$\left(\frac{dp_H}{dv}\right)_o = \left(\frac{dp_s}{dv}\right)_o, \quad \left(\frac{d^2 p_H}{dv^2}\right)_o = \left(\frac{d^2 p_s}{dv^2}\right)_o \tag{1.15a}$$

$$\left(\frac{d^3 p_H}{dv^3}\right)_o = \left(\frac{d^3 p_s}{dv^3}\right)_o - \frac{\Gamma_o}{2v_o}\left(\frac{d^2 p_s}{dv^2}\right)_o \tag{1.15b}$$

$$\left(\frac{d^3p_s}{dv^3}\right)_o = \left(\frac{d^3p_H}{dv^3}\right)_o + \frac{\Gamma_o}{2v_o}\left(\frac{d^2p_H}{dv^2}\right)_o \tag{1.15c}$$

Note that the second term of Eq. (1.15) is a result of the entropy rise which causes shock heating. Also the two curves $p_H(v)$ and $p_s(v)$ are in second-order contact at ($v = v_0$), as indicated by Eq. (1.15a).

For generality we have considered $p_H(v)$ and $p_s(v)$ as arbitrary EOS. Eq. (1.14a) implies (a) $\left(\frac{d^2p_s}{dv^2}\right)_o > 0$ and $v_o > v$ for positvie shocks, namely, compression shocks, and (b) $\left(\frac{d^2p_s}{dv^2}\right)_o < 0$ and $v_o < v$ for negative shocks or rarefaction shocks (see Section 4.2 for more detail). While specific EOS are mostly accurate to the second order, Eqs. (1.15a) and (1.15b) can serve to construct third-order EOS $p_H(v)$ from a given $p_s(v)$ and vice versa provided that $\left(\frac{d^3p}{dv^3}\right)_o$ is as accurate as $\left(\frac{dp}{dv}\right)_o$ and $\left(\frac{d^2p}{dv^2}\right)_o$.

1.7 Shock behavior and properties with an arbitrary EOS

As shown in Sections 1.4 and 1.6, a shock discontinuity is governed by Eqs. (1.1) - (1.7), (1.14a), and (1.14b). The pertinent formulation is applicable to an arbitrary rather than specific EOS as in Chapter 2. Depending upon some properties of the EOS, a SW may be positive (compression) or negative (rarefaction). As dictated by shock formation and stability, it is forbidden that compression shock and rarefaction shock coexist. When compression waves develop into a sharp discontinuity with $d^2p_s/dv^2 > 0$ as required by $s_2 > s_1$ (irreversible adiabatic or shock compression), rarefaction waves can only propagate continuously with $s_2 = s_1$ (reversible adiabatic/isentropic expansion). Such behavior is common with most EOS lacking the critical state $dp_s/dv = d^2p_s/dv^2 = 0$. On the other hand, it is impossible for compression waves to become a shock discontinuity if a rarefaction shock forms with $d^2p_s/dv^2 < 0$ and $s_2 > s_1$. In this connection the compression wave must be continuous with $s_2 = s_1$. Such anomalous behavior is uncommon, except for retrograde fluids with large heat capacity ($c_v/R >> 1$) near the critical state of van der Waals EOS. See Section 4.2. Now it is worthwhile to summarize the fundamental properties of SW as follows.

Positive SW have the change of states with $p_2 > p_1 > 0$ ($h_2 > h_1$), $\rho_2 > \rho_1 > 0$ ($v_1 > v_2 > 0$, $u_2 < u_1$, $E_2 > E_1$), $u_1 > c_1$ ($M_1 > 1$), $u_2 < c_2$ ($M_2 < 1$), and $s_2 > s_1$ for EOS with $dp_s/dv < 0$ and $d^2p_s/dv^2 > 0$ ($c_2 > c_1$, also shock heating $Q > 0$, $T_2 > T_1$). With $u_i = U - \mu_i$ ($i = 1, 2$), we have $U \geq u_1 > u_2$ ($\mu_2 \geq \mu_1 \geq 0$) or simply $U > u_1 > c_2 > u_2 > c_1$. Apparently Fig. 1.2 depicts all these lucidly. From Eqs. (1.7) and (1.8) we have

$$\int_{s_1}^{s_2} Tds = \frac{1}{2}(p_1 + p_2)(v_1 - v_2) - \int_{v_2}^{v_1} pdv = \text{ shaded area } Q > 0.$$

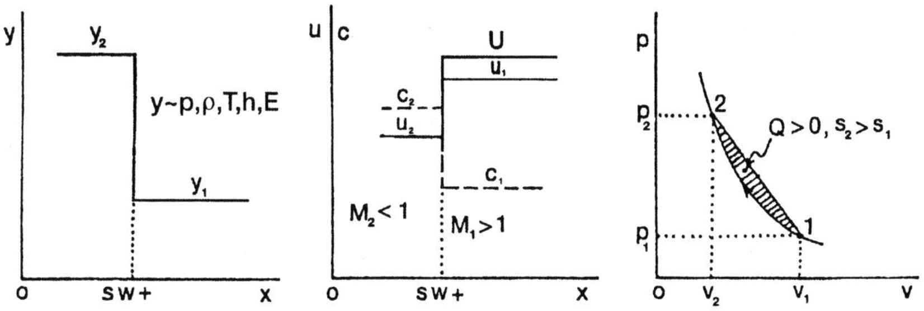

Fig. 1.2 Fundamental properties of compression SW+

Also see Chapter 2 for more specifics with γ-law gas.

On the other hand, negative shocks are featured with $p_1 > p_2 > 0$, $v_2 > v_1 > 0$ ($\rho_1 > \rho_2 > 0$, $u_2 > u_1$, $h_2 < h_1$, $E_2 < E_1$), and $s_2 > s_1$ for EOS with $dp_s/dv < 0$ and $d^2p_s/dv^2 < 0$. Thus the change of states appears just opposite (see Fig. 1.2 and 1.3). Now $c_2 > u_2 > u_1 > c_1$ follows from $M_1 > 1$ and $M_2 < 1$ (for shock stability). The shaded area $Q > 0$ implies $T_1 > T_2$, although T_2 is not far off T_1 because of $c_v/R >> 1$. This and $dc^2/dp <$ 0 suffice to account for $c_2 > c_1$ (note $d^2p_s/dv^2 = \rho^2 \frac{d}{d\rho}(\rho c)^2 = (\rho c)^2 \frac{d}{dp}(\rho c)^2 = \rho^4 c^2 \frac{dc^2}{dp}$ +

$2\rho^3c^2 < 0$ or $\frac{dc^2}{dp} < -\frac{2}{\rho} < 0$). Fig. 1.3 provides a lucid description for all these. See also

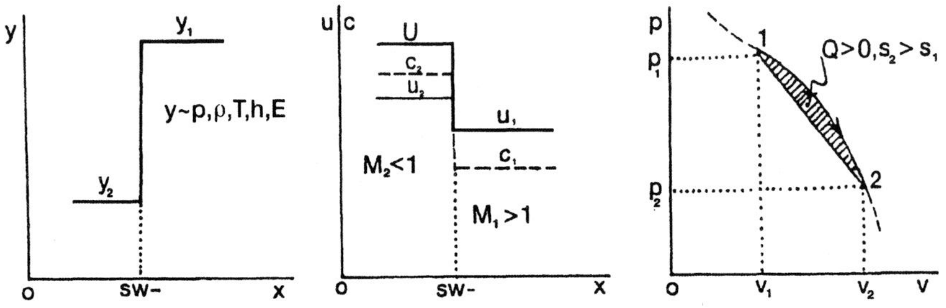

Fig. 1.3 Fundamental properties of rarefaction SW-

Section 4.2 for more specifics about negative SW.

CHAPTER 2 Theory of Shock Waves in γ-Law Gases

2.1 Normal shocks

By normal SW we mean both velocities u_1 and u_2 are perpendicular to the shock front, and all γ-law gases are more or less a replica of the ideal gas with $pv = RT = (\gamma - 1)E$. See Eq. (1.11). Note that R and γ are two specified constants. Substituting $E = pv/(\gamma - 1)$ into Eq. (1.7) we readily obtain

$$\frac{p_2}{p_1} = \frac{(\gamma+1)v_1 - (\gamma-1)v_2}{(\gamma+1)v_2 - (\gamma-1)v_1} \tag{2.1a}$$

or

$$y = \frac{p_2}{p_1} = \frac{\beta x - 1}{\beta - x} \quad \text{with } \beta = \frac{\gamma+1}{\gamma-1} \text{ and } x = \frac{v_1}{v_2} \tag{2.1b}$$

$$\frac{\rho_2}{\rho_1} = \frac{v_1}{v_2} = \frac{(\gamma+1)p_2 + (\gamma-1)p_1}{(\gamma+1)p_1 + (\gamma-1)p_2} \tag{2.2a}$$

or

$$x = \frac{v_1}{v_2} = \frac{\beta y + 1}{\beta + y} \tag{2.2b}$$

These are the useful shock EOS, and Eqs. (2.1b) and (2.2b) will be expediently used later in the analysis of shock reflection. As mentioned earlier, shock waves involve an irreversible change with entropy jump $s_2 > s_1$. That is why the shock adiabat Eq. (2.1a) differs from the isentrope Eq. (1.13), $p_2/p_1 = (v_1/v_2)^\gamma$, which is reversible with $s_2 = s_1$.

From the above results we can express the equations of normal shocks in terms of the Mach number $M = u/c$ with sonic velocity $c^2 = (\partial p/\partial \rho)_s = \gamma p/\rho$. They will prove very useful later. Now that $\rho u^2 = \rho c^2 M^2 = \gamma p M^2$, Eqs. (2.2a), (1.1), and (1.2) can be combined to give

$$\frac{2(p_2 - p_1)}{(\gamma+1)p_1 + (\gamma-1)p_2} = \frac{\rho_2}{\rho_1} - 1 = \frac{u_1}{u_2} - 1 = \frac{p_2 - p_1}{\rho_2 u_2^2}$$

and hence

$$2\rho_2 u_2^2 = (\gamma+1)p_1 + (\gamma-1)p_2$$

Substituting this back into Eq. (1.2) gives

$$p_1(1+\gamma M_1^2) = p_1 + \rho_1 u_1^2 = p_2 + \rho_2 u_2^2 = \tfrac{1}{2}(\gamma+1)(p_2+p_1)$$

and therefore

$$\frac{p_2}{p_1} = 1 + \frac{2\gamma}{\gamma+1}(M_1^2 - 1) \tag{2.3}$$

On the other hand, we have $p_1(1+\gamma M_1^2) = p_2(1+\gamma M_2^2)$ or

$$\frac{1+\gamma M_1^2}{1+\gamma M_2^2} = \frac{p_2}{p_1} = 1 + \frac{2\gamma}{\gamma-1}(M_1^2 - 1)$$

Thus we deduce

$$M_2^2 = \frac{(\gamma-1)M_1^2 + 2}{2\gamma M_1^2 - (\gamma-1)} \tag{2.4}$$

Further manipulations yield

$$\frac{\rho_2}{\rho_1} = \frac{(\gamma+1)M_1^2}{(\gamma-1)M_1^2 + 2} \tag{2.5}$$

$$\frac{T_2}{T_1} = \frac{[2\gamma M_1^2 - (\gamma-1)][(\gamma-1)M_1^2 + 2]}{(\gamma+1)^2 M_1^2} \tag{2.6}$$

and

$$\frac{s_2 - s_1}{c_v} = \ell n\left[\frac{p_2}{p_1}\left(\frac{\rho_1}{\rho_2}\right)^{\gamma}\right] = \ell n\left[1 + \frac{2\gamma}{\gamma+1}(M_1^2 - 1)\right] + \gamma \ell n\left[1 - \frac{2(M_1^2 - 1)}{(\gamma+1)M_1^2}\right]$$

$$\approx \frac{2\gamma(\gamma-1)}{3(\gamma+1)^2}\left(M_1^2-1\right)^3 = \frac{\gamma^2-1}{12\gamma^2}\left(\frac{p_2}{p_1}-1\right)^3 \tag{2.7}$$

Note that the last line of Eq. (2.7) is a result of logarithmic-series expansion to the third order only. This can be deduced otherwise from Eqs. (1.14b), (1.13), (1.15a), and (2.3), viz.,

$$s_2 - s_1 = \frac{(\gamma+1)v_1(p_2-p_1)^3}{12T_1\gamma^2 p_1^2} = \frac{(\gamma+1)R}{12\gamma^2}\left(\frac{p_2}{p_1}-1\right)^3 = \frac{(\gamma^2-1)c_v}{12\gamma^2}\left(\frac{p_2}{p_1}-1\right)^3$$

$$= \frac{2\gamma(\gamma-1)c_v}{3(\gamma+1)^2}\left(M_1^2-1\right)^3.$$

For a shock wave moving into a gas at rest, we may write u_1 = U and u_2 = U - u with gas velocity u behind the shock front. From Eqs. (1.2) and (2.3), we deduce

$$\rho_1 u U = p_2 - p_1 = \frac{2\gamma p_1}{\gamma+1}\left(M_1^2-1\right) = \frac{2\rho_1 c_1^2}{\gamma+1}\left(\frac{U^2}{c_1^2}-1\right)$$

and hence

$$U^2 = c_1^2 + \tfrac{1}{2}(\gamma+1)uU \tag{2.8}$$

which is known as the shock Hugoniot. Note that Eqs. (2.1a), (2.3), and (2.8) are essentially three versions of the shock wave EOS from which other thermodynamic EOS can be sought. Returning to Eqs. (2.3) to (2.8), we note in the strong shock limit $M_1^2 = \frac{U^2}{c_1^2} = \frac{\rho_1 U^2}{\gamma p_1} >> 1$ and hence

$$\frac{p_2}{p_1} = \frac{2\gamma}{\gamma+1}M_1^2 \tag{2.3a}$$

$$p_2 = \frac{2\rho_1 U^2}{\gamma+1} \tag{2.3b}$$

$$\frac{\rho_2}{\rho_1} = \frac{\gamma+1}{\gamma-1} = \beta \tag{2.5a}$$

$$\frac{T_2}{T_1} = \frac{2\gamma(\gamma-1)}{(\gamma+1)^2} M_1^2 \tag{2.6a}$$

$$T_2 = \frac{2U^2}{(\gamma+1)^2 c_v} \tag{2.6b}$$

$$M_2^2 = \frac{\gamma-1}{2\gamma} < 1 \text{ (subsonic)} \tag{2.4a}$$

$$\frac{\Delta s}{c_v} = \ell n\left[\frac{2\gamma}{\gamma+1}\left(\frac{\gamma-1}{\gamma+1}\right)^{\gamma} M_1^2\right] > 0 \tag{2.7a}$$

$$U = \tfrac{1}{2}(\gamma+1)u \tag{2.8a}$$

On the other hand, weak shocks degenerate into sound waves with $U \approx c_1$. It is worth noting that Eqs. (2.3a) - (2.8a) delineate the salient properties of strong shocks, which are very useful in the theory of shock waves. When γ is interpreted as a variety of EOS constants, the foregoing equations can serve as a building block of all SW studies today. They are adoptable to many shock environments by modification, generalization, or analogy.

2.2 Oblique Shocks

Now let us take a step forward to examine the analytical aspects of oblique shocks. By obliquity we mean that velocity vector w_1 intersects the shock front with a shock angle θ and velocity vector w_2 deviates from the direction of w_1 with a deflection angle of δ. The velocity vector w_i may be resolved into its normal and tangential components, viz. $w_i = \boldsymbol{u}_i + \boldsymbol{b}_i$ with i = 1, 2 as sketched in Fig. 2.1(b). The deflection δ

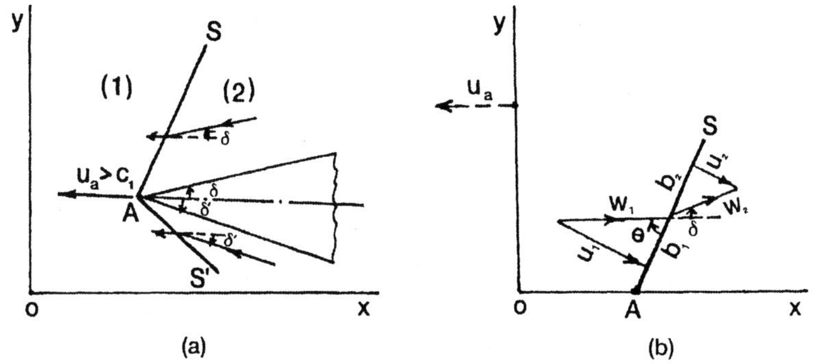

Fig. 2.1 Oblique SW: (a) formation of AS and AS′ at apex of moving wedge in laboratory coordinates, (b) geometry of AS in moving coordinates.

is apparently due to the frictionless slip $\boldsymbol{b_1} = \boldsymbol{b_2}$ and the normal jump $u_2 < u_1$. From Fig. 2.1(b) we may formulate the equations of an inert oblique shock as follows.

$$\rho_1 w_1 \sin\theta = \rho_2 w_2 \sin(\theta - \delta) \tag{2.9}$$

$$w_1 \cos\theta = w_2 \cos(\theta - \delta) \tag{2.10}$$

$$p_1 + \rho_1 w_1^2 \sin^2\theta = p_2 + \rho_2 w_2^2 \sin^2(\theta - \delta) \tag{2.11}$$

$$h_1 + \tfrac{1}{2} w_1^2 \sin^2\theta = h_2 + \tfrac{1}{2} w_2^2 \sin(\theta - \delta) \tag{2.12}$$

It is interesting to note that Eqs. (2.9), (2.11), (2.12) and $h = \gamma p v/(\gamma - 1)$ can be combined to give

$$h_2 - h_1 = \tfrac{1}{2}(v_1 + v_2)(p_2 - p_1)$$

and

$$\frac{\rho_2}{\rho_1} = \frac{(\gamma+1)p_2 + (\gamma-1)p_1}{(\gamma-1)p_2 + (\gamma+1)p_1}$$

which are the same as Eqs. (1.6) and (2.2a). Thus the Rankine-Hugoniot equation and shock EOS are unique in all shock environments. Now that we have $M_1 = w_1/c_1$ or $\rho_1 w_1^2 = \gamma p_1 M_1^2$, Eqs. (2.9) and (2.11) can be combined to give

$$\frac{p_2}{p_1} = 1 + \frac{2\gamma}{\gamma+1}\left(M_1^2 \sin^2\theta - 1\right) \tag{2.13}$$

$$\frac{\rho_2}{\rho_1} = \frac{(\gamma+1)M_1^2 \sin^2\theta}{(\gamma-1)M_1^2 \sin^2\theta + 2} \tag{2.14}$$

$$\frac{T_2}{T_1} = \frac{\left(2\gamma M_1^2 \sin^2\theta - \gamma + 1\right)\left[2 + (\gamma-1)M_1^2 \sin^2\theta\right]}{(\gamma+1)^2 M_1^2 \sin^2\theta} \tag{2.15}$$

$$\frac{s_2 - s_1}{c_v} = \ell n\left[1 + \frac{2\gamma}{\gamma+1}\left(M_1^2 \sin^2\theta - 1\right)\right] + \gamma \ell n\left[1 - \frac{2\left(M_1^2 \sin^2\theta - 1\right)}{(\gamma+1)M_1^2 \sin^2\theta}\right] \tag{2.16}$$

which turn out to be a slight modification of Eqs. (2.3), (2.5), (2.6), and (2.7) respectively. These equations hold good for arc $\sin(1/M_1) \le \theta \le 90°$, with $\sin\theta = 1/M_1$ for the acoustic perturbation and with $\theta = 90°$ (also $\delta = 0$) for the normal shock exactly. Manipulations with Eqs. (2.11) and (2.13) yield

$$M_2^2 \sin^2(\theta - \delta) = \frac{(\gamma-1)M_1^2 \sin^2\theta + 2}{2\gamma M_1^2 \sin^2\theta - \gamma + 1} \tag{2.17a}$$

which is equivalent to Eq. (2.4) with the obliquity and deflection taken into account.

It should be noted that the deflection angle δ has an important connection hereafter. From Eqs. (2.9) and (2.10), we may write

$$r = \frac{\rho_2}{\rho_1} = \frac{w_1 \sin\theta}{w_2 \sin(\theta - \delta)} = \frac{\tan\theta}{\tan(\theta - \delta)} = \frac{1 + \tan\theta \tan\delta}{1 - \cot\theta \tan\delta}$$

and

$$\tan\delta = \frac{r - 1}{\tan\theta + r\cot\theta} = \frac{r - 1}{\tan\theta\left(1 + r\cot^2\theta\right)}$$

From Eq. (2.14), now we have

$$r - 1 = \frac{2\left(M_1^2 \sin^2\theta - 1\right)}{(\gamma - 1)M_1^2 \sin^2\theta + 2}$$

and

$$1 + r\cot^2\theta = 1 + \frac{(\gamma + 1)M_1^2 \cos^2\theta}{(\gamma - 1)M_1^2 \sin^2\theta + 2} = \frac{M_1^2(\gamma + \cos 2\theta) + 2}{(\gamma - 1)M_1^2 \sin^2\theta + 2}$$

Accordingly we obtain

$$\tan\delta = \frac{2\left(M_1^2 \sin^2\theta - 1\right)}{\tan\theta\left[(\gamma + 1)M_1^2 - 2\left(M_1^2 \sin^2\theta - 1\right)\right]} = \frac{M_1^2 \sin 2\theta - 2\cot\theta}{M_1^2 \cos 2\theta + \gamma M_1^2 + 2} \tag{2.18}$$

by direct substitution of the r-terms. Again, Eqs. (2.9) and (2.10) may be combined to give $\cot(\theta-\delta) = (\rho_2/\rho_1)\cot\theta$. Then Eq. (2.17a) becomes

$$M_2^2 = \frac{(\gamma - 1)M_1^2 \sin^2\theta + 2}{2\gamma M_1^2 \sin^2\theta - \gamma + 1}\left[1 + \left(\frac{\rho_2}{\rho_1}\cot\theta\right)^2\right]$$

because of $\csc^2(\theta-\delta) = 1 + \cot^2(\theta-\delta)$. Substituting Eq. (2.14) herein gives

$$M_2^2 = \frac{(\gamma + 1)^2 M_1^4 \sin^2\theta - 4\left(M_1^2 \sin^2\theta - 1\right)\left(\gamma M_1^2 \sin^2\theta + 1\right)}{\left(2\gamma M_1^2 \sin^2\theta - \gamma + 1\right)\left[(\gamma - 1)M_1^2 \sin^2\theta + 2\right]} \tag{2.17b}$$

It will become clear that Eqs. (2.17b) and (2.18) are very useful in the analysis of shock reflection and diffraction. Given M_1, Eq. (2.18) can be plotted as a θ vs δ shock polar (see Fig. 2.2) which has two critical points (δ_x, θ_x) and (δ_s, θ_s). Let us re-write Eq. (2.18) as

$$\cot\delta = \sqrt{\frac{x}{1 - x}}\left(\frac{A}{Bx - 1} - 1\right)$$

with $x = \sin^2\theta$, $A = \frac{\gamma+1}{2}M_1^2$, and $B = M_1^2$. Now that $\frac{d\delta}{dx} = 0$ requires $B(B-2A)x^2 + B(A-2)x$ $+(A+1) = (-\gamma M_1^4)x^2 + 2M_1^2\left(\frac{\gamma+1}{4}M_1^2 - 1\right)x + \left(1 + \frac{\gamma+1}{2}M_1^2\right) = 0$, solving this for x gives

$$\sin^2\theta_x = \frac{1}{\gamma M_1^2}\left\{\frac{\gamma+1}{4}M_1^2 - 1 + \left[(\gamma+1)\left(1 + \frac{\gamma-1}{2}M_1^2 + \frac{\gamma+1}{16}M_1^4\right)\right]^{\frac{1}{2}}\right\} \tag{2.19}$$

and

$$\cot\delta_x = \sqrt{\frac{\sin^2\theta_x}{1-\sin^2\theta_x}}\left[\frac{(\gamma+1)M_1^2}{2\left(M_1^2\sin^2\theta_x - 1\right)} - 1\right] \tag{2.20}$$

Thus we have the particular point (δ_x, θ_x) for the maximum deflection of an oblique shock, which leads to a detached bow shock or Mach reflection. Let $M_2 = 1$ in Eq. (2.17b) which gives

$$\sin^2\theta_s = \frac{1}{\gamma M_1^2}\left\{\frac{\gamma+1}{4}M_1^2 - \left(\frac{3-\gamma}{4}\right) + \left[(\gamma+1)\left(\frac{9+\gamma}{16} - \frac{3-\gamma}{8}M_1^2 + \frac{\gamma+1}{16}M_1^4\right)\right]^{\frac{1}{2}}\right\} \tag{2.21}$$

Replacing θ_x by θ_s in Eq. (2.20) gives $\cot\delta_s$, and hence we obtain the sonic transition point (δ_s, θ_s) of the θ vs δ shock polar. It may be noted that the maximum deflection δ_x and the sonic deflection δ_s (at $M_2 = 1$) are very close to each other by virtue of Eqs. (2.19) and (2.21). For $M_1^2 \gg 1$, we have $\theta_x = \theta_s = \arcsin\sqrt{\frac{\gamma+1}{2\gamma}}$ and $\delta_x = \delta_s = \text{arc}\cot\sqrt{\gamma^2 - 1}$.

In aerodynamics it is often convenient to refer to (a) subsonic flows with $M < 1$, (b) transonic flows with $0.8 \leq M \leq 1.2$, (c) supersonic flows with $1.2 \leq M \leq 5$, and (d) hypersonic flows with $5 \leq M \leq 40$. By transonic approximation, the deflection angles are simply given by

$$\delta_x = \frac{4\left(M_1^2 - 1\right)^{\frac{3}{2}}}{3\sqrt{3}(\gamma+1)M_1^2} \tag{2.22}$$

and

$$\delta_s = \frac{\left(M_1^2 - 1\right)^{\frac{3}{2}}}{\sqrt{2}(\gamma+1)M_1^2} \tag{2.23}$$

On the other hand, the strong oblique shock limit is featured by $M_1^2 \sin^2\theta >> 1$ and

$$\frac{p_2}{p_1} = \frac{2\gamma}{\gamma+1} M_1^2 \sin^2\theta \tag{2.13a}$$

$$\frac{\rho_2}{\rho_1} = \frac{\gamma+1}{\gamma-1} = \beta \tag{2.14a}$$

$$\frac{T_2}{T_1} = \frac{2\gamma(\gamma-1)}{(\gamma+1)^2} M_1^2 \sin^2\theta \tag{2.15a}$$

$$\frac{\Delta s}{c_v} = \ell n \left[\left(\frac{\gamma-1}{\gamma+1}\right)^{\gamma} \left(\frac{2\gamma}{\gamma+1} M_1^2 \sin^2\theta\right) \right] \tag{2.16a}$$

$$\tan\delta = \frac{\sin 2\theta}{\cos 2\theta + \gamma} \approx \frac{\sin 2\theta}{\gamma+1} \tag{2.18a}$$

$$M_2^2 = \frac{(\gamma+1)^2 \csc^2\theta - 4\gamma}{2\gamma(\gamma-1)} \tag{2.17c}$$

For small δ and θ, Eq. (2.18) may be written as

$$\delta = \frac{2}{\theta}\left[\frac{M_1^2\theta^2 - 1}{(\gamma+1)M_1^2 - 2\left(M_1^2\theta^2 - 1\right)}\right] \approx \frac{2\theta}{(\gamma+1) - 2\theta^2}$$

which is essentially the same as Eq. (2.18a) by dropping out the negligible $2\theta^2$.

Further notes should be taken of the general properties of oblique shocks. As sketched in Fig 2.2, Eq. (2.18) implies that each deflection angle δ is associated with two shock angles ($\theta_v > \theta_w$, v being the vigorous or strong and w the weak solution)

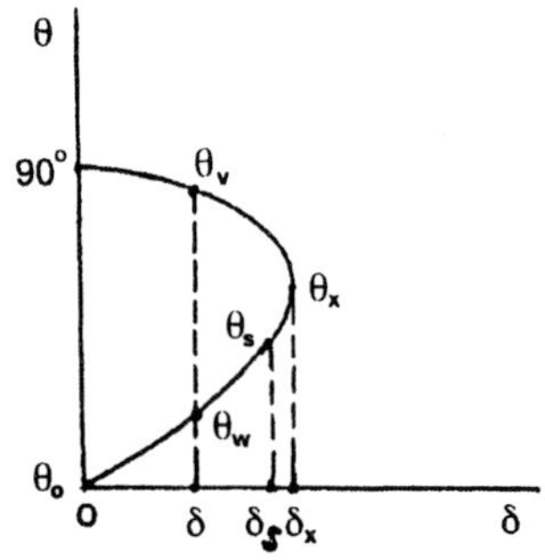

Fig 2.2 Shock polar θ vs δ.

except for the double-root solution (δ_x, θ_x) of $\theta_v = \theta_w = \theta_x$. Yet another special solution of Eq. (2.18) is $\delta = 0$ for both $\theta_o = \arcsin(1/M_1)$ and $\theta = 90°$. Now the shock polar is divided by the point (δ_s, θ_s) into two branches, viz. the supersonic branch $\theta_o - \theta_w - \theta_s$ with M_2 varying from $M_2 > 1$ (i.e. $M_2 = M_1$ at $(0, \theta_o)$ by Eq. (2.17a)) to $M_2 = 1$ at (δ_s, θ_s) and the subsonic branch $\theta_s - \theta_x - \theta_v - 90°$ with M_2 varying from $M_2 = 1$ to $M_2 < 1$. As mentioned earlier, M_1 is constant along the whole shock polar (also $M_1 > 1$ in view of $\sin\theta_o = 1/M_1 < 1$). Thus weak oblique shocks are supersonic in the sense of $M_1 > M_2 > 1$, and strong oblique shocks are subsonic with $\theta_v > \theta_w$, $M_1 > 1$, and $M_2 \leq (\gamma-1)/2\gamma < 1$. Moreover, for $\theta = 90°$ and $\delta = 0°$, oblique shocks become normal SW (see Eqs. (2.9) - (2.17) versus Eq. (2.3) - (2.7), Eqs. (2.13a) - (2.17c) versus Eqs. (2.3a) - (2.7a), and also Fig 2.1). Normal shocks are stronger than all oblique shocks.

2.3 Spherical and cylindrical blast waves

Air burst of chemical and nuclear explosions sends out blast waves in spherical or cylindrical forms if the sources may be treated as point or line charges respectively. Rigorous analysis of these is carried out elsewhere by the method of self-similar flow or by computer simulation. In order to gain a basic understanding, we seek to provide a simplified model for the point explosion as a shock sphere in Fig. 2.3.

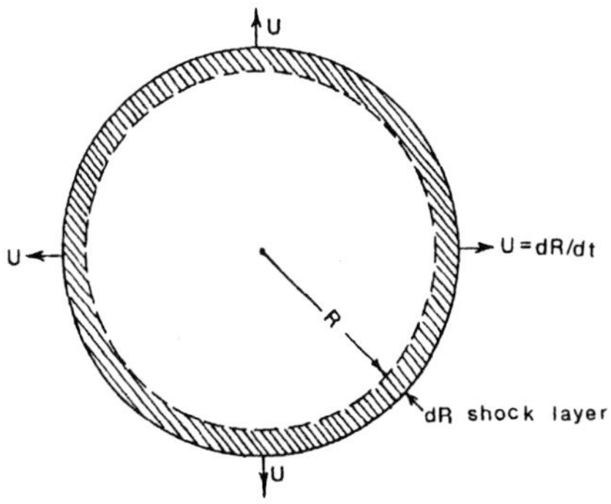

Fig. 2.3 Shock sphere due to point explosion.

When a point explosion suddenly releases a large amount of energy Y in the open air, a shock sphere of radius R is formed with mass $M = 4\pi R^3 \rho_1/3$ in the state (p_1, ρ_1). The near-field effects of a strong shock create a thin layer of dense gas with $M = 4\pi R^2 (dR)\rho_2$ in the state (p_2, ρ_2). From the conservation of mass, we infer

$$\frac{dR}{R} = \frac{\rho_1}{3\rho_2} = \frac{1}{3}\left(\frac{\gamma - 1}{\gamma + 1}\right) \qquad (2.24)$$

where Eq. (2.5a) has been used. For $\gamma = 1.4$ we have $dR/R = 1/18$ which validates the shock discontinuity of $dR << R$. Once the spherical shock front starts to advance with a velocity $U = dR/dt$, we may write

$$4\pi R^2 p_1 = F = \frac{d}{dt}(Mu) = \frac{4}{3}\pi\rho_1 \frac{d}{dt}\left(R^3 u\right)$$

Let $p_1 = \alpha p_2$, α being an important constant of this analysis. The above equation of motion becomes

$$3\alpha R^2 p_2 = \rho_1 \frac{d}{dt}\left(R^3 u\right) = \rho_1 U \frac{d}{dR}\left(R^3 u\right)$$

Substituting Eqs. (2.3b) and (2.8a) herein yields

$$3(\alpha - 1)R^2 U dR = R^3 dU$$

which can be integrated as

$$U = BR^{3(\alpha-1)} \tag{2.25}$$

with another constant B to be determined soon.

Now the conservation of energy may be written as

$$Y = \frac{1}{2}Mu^2 + \frac{p_1 V}{\gamma - 1} = \frac{8}{3}\pi\rho_1 U^2 R^3\left[\frac{1}{(\gamma+1)^2} + \frac{\alpha}{\gamma^2 - 1}\right] = \frac{8}{3}\pi\rho_1\left[\frac{1}{(\gamma+1)^2} + \frac{\alpha}{\gamma^2 - 1}\right]B^2 R^{3(2\alpha-1)}$$

which requires 2α - 1 = 0 for the equation and Y to be independent of time. Accordingly we deduce α = 1/2 and

$$B = \sqrt{\frac{3Y(\gamma - 1)(\gamma + 1)^2}{4\pi\rho_1(3\gamma - 1)}} \tag{2.26}$$

It may be noted that Eqs. (2.24) - (2.26) are the results of mass, momentum, and energy conservation. Substituting $\alpha = \frac{1}{2}$ in Eq. (2.25) we have $\frac{dR}{dt} = U = BR^{-\frac{2}{3}}$ and hence

$$R = \left(\frac{5}{2}B\right)^{\frac{2}{5}} t^{\frac{2}{5}} = \left(\frac{75Y(\gamma - 1)(\gamma + 1)^2}{16\pi\rho_1(3\gamma - 1)}\right)^{\frac{1}{5}} t^{\frac{2}{5}} \tag{2.27}$$

where t is the shock travel time. This is the equation of shock trajectory. From Eqs. (2.3b), (2.25) - (2.27), we obtain

$$p_2 = \frac{2\rho_1 B^2}{(\gamma + 1)R^3} = \frac{2\rho_1\left(\frac{2}{5}\right)^{\frac{6}{5}} B^{\frac{4}{5}}}{(\gamma + 1)t^{\frac{6}{5}}}$$

and hence

$$p_2 R^3 = \frac{3Y}{2\pi}\left(\frac{\gamma^2-1}{3\gamma-1}\right) = \text{constant} \tag{2.28}$$

$$p_2 t^{\frac{6}{5}} = \left\{\frac{24Y}{125\pi}\left(\frac{\gamma-1}{3\gamma-1}\right)\sqrt{\frac{2\rho_1^3}{\gamma+1}}\right\}^{\frac{2}{5}} = \text{constant} \tag{2.29}$$

The last two equations not only indicate the shock decay due to the spherical divergence but also serve as the basis of the cube-root scaling laws of blast waves: $\frac{R}{R_1} = \left(\frac{Y}{Y_1}\right)^{\frac{1}{3}} = \frac{t}{t_1}$ in view of $pR^3 \sim Y$ and $pt^{\frac{6}{5}} \sim Y^{\frac{2}{5}}$ from Eqs. (2.28) and (2.29).

The foregoing analysis can be carried over by replacing $\pi R^2 l \rho_1 = M = 2\pi R l (dR) \rho_2$ for cylindrical blast waves (here l being the length of line charge Y). The analytical results are as follows.

$$\frac{dR}{R} = \frac{\rho_1}{2\rho_2} = \frac{1}{2}\left(\frac{\gamma-1}{\gamma+1}\right) = \frac{1}{12} \text{ for } \gamma = 1.4 \tag{2.24a}$$

$$U = BR^{2(\alpha-1)} = B/R \text{ for } \alpha = 1/2 \tag{2.25a}$$

$$Y = 2\pi\rho_1 l\left[\frac{1}{(\gamma+1)^2} + \frac{\alpha}{\gamma^2-1}\right]B^2 R^{2(2\alpha-1)}$$

$$B = \sqrt{\frac{Y(\gamma-1)(\gamma+1)^2}{\pi\rho_1 \ell(3\gamma-1)}} \tag{2.26a}$$

$$R = \sqrt{2Bt} \tag{2.27a}$$

$$p_2 R^2 = \frac{2Y}{\pi l}\left(\frac{\gamma^2-1}{3\gamma-1}\right) = \text{constant} \tag{2.28a}$$

$$p_2 t = \sqrt{\frac{Y\rho_1}{\pi l}\left(\frac{\gamma-1}{3\gamma-1}\right)} = \text{constant} \tag{2.29a}$$

As mentioned earlier, all these results highlight the properties of the blast waves at the very early stage only. Thus a nuclear burst sets off a 10^7°K fireball of radioactive products, whose high pressures form a shock front of tens or hundreds of Mb (million atmospheres = Mb). Such a strong SW is immediately transmitted to the ambient atmosphere as a blast wave. The aftermath of these events is too complicated to be included in this inquiry.

2.4 Normal shock reflection and transmission

When a shock wave encounters some obstacle, the interaction results will be a reflected shock or rarefaction wave, a transmitted shock, and diffracted shocks if the obstacle causes the incident shock to change its path. But shock propagation and interaction are not so easily understandable as the linear waves of acoustics and optics. Now Fermat principle, Snell's laws, Fresnel diffraction, and Fraunhofer diffraction are all not applicable here. So we have to take a non-linear wave approach for the inquiry.

Let us consider the normal shock reflection at a rigid wall. Because the gas must be at rest near the wall, we will denote this state (0) by (p_0, ρ_0, u_0 = U_i, h_0) for the incident normal shock S_i and state (2) by (p_2, ρ_2, u_2 = U_r, h_2) for the reflected shock S_r. State (1), (p_1, ρ_1, u_1, h_1) is common to both S_i and S_r, but we have $U_i - u = u_1 = U_r + u$ because of the reversal of shock flow (here u = the gas velocity behind S_i but ahead of S_r). Thus our shock reflection problem is formulated as follows.

For the incident shock S_i

$$\rho_0 U_i = \rho_1 (U_i - u)$$

$$p_1 - p_0 = \rho_0 u U_i$$

$$x_i = \frac{y_i + \beta}{1 + \beta y_i} \text{ with } x_i = \frac{\rho_o}{\rho_1} = \frac{v_1}{v_o},\ y_i = \frac{p_1}{p_o},\ \text{and } \beta = \frac{\gamma + 1}{\gamma - 1}$$

For the reflected shock S_r

$$\rho_2 U_r = \rho_1 (U_r + u)$$

$$p_2 - p_1 = \rho_2 u U_r$$

$$x_r = \frac{y_r + \beta}{1 + \beta y_r} \text{ with } x_r = \frac{\rho_1}{\rho_2} = \frac{v_2}{v_1},\ y_r = \frac{p_2}{p_1}$$

These equations are all given earlier. See Eqs. (1.1), (1.2), and (2.2b). From the above two sets of continuity and momentum equations, we deduce

$$(p_2 - p_1)(v_1 - v_2) = u^2 = (p_1 - p_o)(v_o - v_1)$$

or

$$y_i x_i (y_r - 1)(1 - x_r) = (y_i - 1)(1 - x_i)$$

Further simplification of this with $z = y_i y_r$ gives

$$\frac{(y_i - 1)^2}{y_i + \beta} = y_i \frac{(y_r - 1)^2}{1 + \beta y_r} = \frac{(z - y_i)^2}{y_i + \beta z}$$

multiplying and factoring out, we have

$$(1 - z)\left[(\beta + 2)y_i^2 - (1 + z)y_i - \beta z\right] = 0$$

which leads to

$$(\beta + 2)y_i - 1 = (\beta + y_i)y_r$$

Then we obtain the equations of normal shock reflection at a rigid wall:

$$\frac{p_2}{p_1} = y_r = \frac{(\beta + 2)y_i - 1}{\beta + y_i} = \frac{(3\gamma - 1)p_1 - (\gamma - 1)p_o}{(\gamma - 1)p_1 + (\gamma + 1)p_o} \tag{2.30}$$

$$\frac{\rho_2}{\rho_1} = \frac{1}{x_r} = \frac{1+\beta y_r}{y_r+\beta} = \frac{(\beta+1)y_r}{2y_r+\beta-1} = \frac{\gamma p_1}{(\gamma-1)p_1+p_o} \qquad (2.31)$$

$$\frac{T_2}{T_1} = y_r x_r = \left(\frac{\gamma-1}{\gamma}+\frac{p_o}{\gamma p_1}\right)\left[\frac{(3\gamma-1)p_1-(\gamma-1)p_o}{(\gamma-1)p_1+(\gamma+1)p_o}\right] = \frac{E_2}{E_1} \qquad (2.32)$$

The strong shock limit is simply

$$\frac{p_2}{p_1} = \frac{3\gamma-1}{\gamma-1} \qquad (2.30a)$$

$$\frac{\rho_2}{\rho_1} = \frac{\gamma}{\gamma-1} \qquad (2.31a)$$

$$\frac{E_2}{E_1} = \frac{T_2}{T_1} = \frac{3\gamma-1}{\gamma} \qquad (2.31a)$$

providing $p_1 >> p_o$ and $\rho_1/\rho_o = (\gamma+1)/(\gamma-1) = \beta$. Thus $p_2/p_1 = 8$, $\rho_2/\rho_1 = 3.5$, $T_2/T_1 = 2.3$, and $\rho_2/\rho_o = 21$ with $\beta = 6$ and $\gamma = 1.4$. These are also the results for the collision of two equal normal/planar shocks (the rigid wall is equivalent to the plane of collision by symmetry). Converging/imploding spherical and cylindrical shocks exhibit similar results:

Converging Shocks / Strength	Planar	Cylindrical	Spherical
	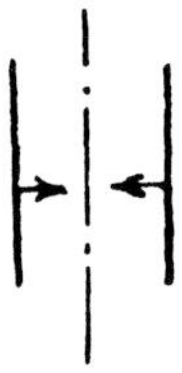	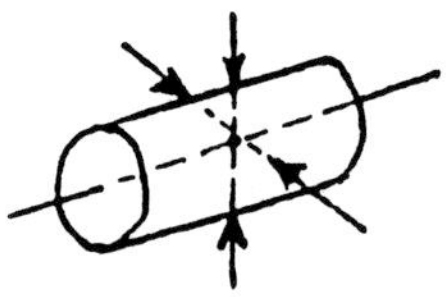	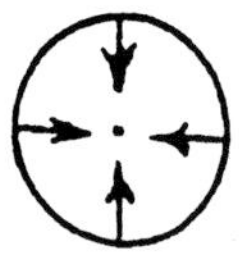
p_2/p_1	8	14	23
ρ_2/ρ_1	3.5	12	23
T_2/T_1	2.3	1.7	1

Note that imploding spherical and cylindrical shocks are unstable with secondary diverging shocks to recur.

If a normal shock advances from one gas (ρ_o, γ, β) to another gas (ρ_o', γ', β'), the interface may cause two modes of shock reflection and transmission. Like acoustic impedance, shock impedances $\rho_o U_i$ and $\rho_o' U_t$ decide the outcome of interaction. Thus we may have (i) a step-up combination of incident S_i (p_1, U_i), reflected S_r (p_2, U_r) and transmitted S_t (p_2', U_t) for shock impedance $\rho_o U_i < \rho_o' U_t$ and $p_2' > p_1$, or (ii) a step-down shock transmission for $\rho_o U_i > \rho_o' U_t$ and $p_2' < p_1$, with a reflected rarefaction wave R_r to diminish the incident shock front. The complete analysis of these is given as follows.

(i) Fig. 2.4 shows S_i (p_1, U_i), S_r (p_2, U_r), S_t (p_2', U_t), and C ($p_2 = p_2'$, $\mu_2 = \mu_2'$) due to $\rho_o U_i < \rho_o' U_t$, where C is a contact discontinuity or interface with density $\rho_2 \neq \rho_2'$ and entropy $S_2 \neq S_2'$. Note both gases initially at rest $\mu_o = \mu_o' = 0$, $p_o = p_o' \neq 0$, $\rho_o \neq \rho_o'$, and

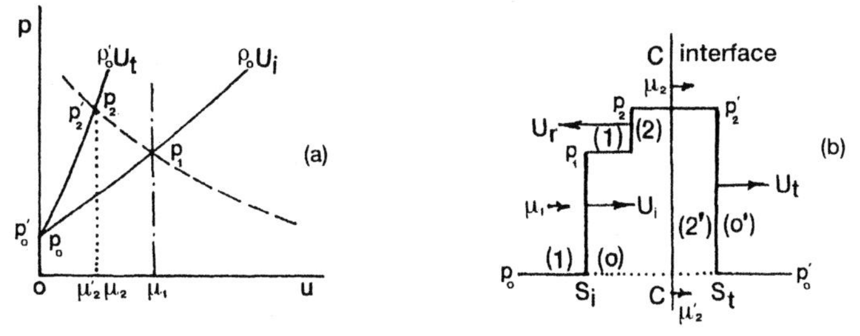

Fig. 2.4 Shock reflection and transmission: (a) impedance mismatch $\rho_o U_i < \rho_o U_t$, (b) step-up shocking $p_2 > p_1$.

$\gamma \neq \gamma'$ (μ_j = gas velocities in states j = 0, 1, 2, 2′ and $u_j = U_k - \mu_j$ = relative velocities for $k = i, r, t$). As before, we may write the essential equations for the incident S_i (p_1, U_i):

$$\rho_o U_i = \rho_1 (U_i - \mu_1)$$

$$p_1 - p_o = \rho_o U_i \mu_1$$

$$x_i = \frac{\rho_o}{\rho_1} = \frac{y_i + \beta}{1 + \beta y_i} \text{ with } y_i = \frac{p_1}{p_o},\ \beta = \frac{\gamma + 1}{\gamma - 1}$$

$$\therefore \mu_1^2 = (p_1 - p_o)(v_o - v_1) = p_o v_o (\beta - 1)(y_i - 1)^2 / (1 + \beta y_i) \tag{2.32i}$$

$$(\rho v = 1,\ v_1 < v_o,\ p_1 > p_o)$$

For the reflected shock S_r (p_2, U_r), we have

$$\rho_1(U_r + \mu_1) = \rho_2(U_r + \mu_2) = m$$

$$p_2 - p_1 = m(\mu_1 - \mu_2) = m^2(v_1 - v_2),\ \ v_2 < v_1,\ \ p_2 > p_1$$

$$x_r = \frac{\rho_1}{\rho_2} = \frac{y_r + \beta}{1 + \beta y_r} \text{ with } y_r = \frac{p_2}{p_1}$$

$$\therefore (\mu_1 - \mu_2)^2 = (p_2 - p_1)(v_1 - v_2) = p_1 v_1 (\beta - 1)(y_r - 1)^2 / (1 + \beta y_r) \tag{2.32r}$$

For the transmitted S_t (p_2', U_t), we have

$$\rho_o' U_t = \rho_2'(U_t - \mu_2') \quad \text{note } \mu_2' = \mu_2 = \text{velocity of interface displaced}$$

$$p_2' - p_o' = \rho_o' U_t \mu_2' \qquad \text{(as mentioned } p_2' = p_2,\ p_o' = p_o \neq 0\text{)}$$

$$x_t = \frac{\rho_0'}{\rho_2'} = \frac{y_t + \beta'}{1 + \beta' y_t} \qquad \text{with } y_t = \frac{p_2'}{p_o'} = y_r y_i \text{ and } \beta' = \frac{\gamma' + 1}{\gamma' - 1}$$

$$\therefore (\mu_2')^2 = (p_2' - p_o')(v_o' - v_2') = p_o' v_o' (\beta' - 1)(y_t - 1)^2 / (1 + \beta' y_t) \tag{2.32t}$$

Now that the impedance mismatch results in $\mu_1 > \mu_2 = \mu_2'$ and $p_1 < p_2' = p_2$, we may write $\mu_1 = \mu_2' + \sqrt{(\mu_1 - \mu_2)^2}$ [see Fig. 2.5(a)]. This combines Eqs. (2.32i), (2.32r), and (3.32t) to give

$$\frac{y_i - 1}{\sqrt{1 + \beta y_i}} = \sqrt{\frac{\rho_o}{\rho_o'}\left(\frac{\beta' - 1}{\beta - 1}\right)} \frac{y_t - 1}{\sqrt{1 + \beta' y_t}} + \sqrt{\frac{y_i + \beta}{1 + \beta y_i} \frac{y_t - y_i}{y_i + \beta y_t}} \tag{2.32}$$

which is solvable by numerical iteration for y_t and $y_r = y_t / y_i$ in terms of the incident shock strength y_i. Henceforth other shock parameters can be sought with no difficulty.

(ii) S_i (p_1, U_i), R_r (p_2, dx/dt), S_t (p_2', U_t), C ($p_2 = p_2'$, $\mu_2 = \mu_2'$) due to $\rho_o U_i \cdot \rho_o' U_t$ with $\mu_1 < \mu_2' = \mu_2$ and $p_1 > p_2' = p_2$ (Fig 2.5). Note that the reflected rarefaction R_r is a

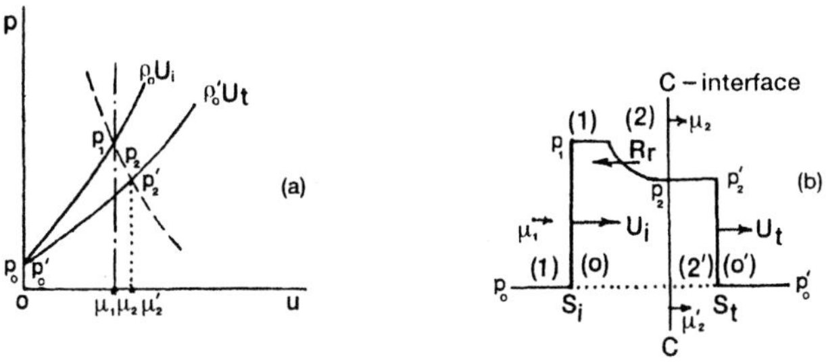

Fig. 2.5 Shock reflection and refraction: (a) impedance mismatch $\rho_o U_i > \rho_o U_t$, (b) step-down transmission $p_2 < p_1$.

non-linear expansion wave with $s_2 = s_1 \neq s_2'$. According to the method of characteristics, R_r propagates against S_i continuously with a velocity $dx/dt = \mu - c$, $\mu_2 + \frac{2c_2}{\gamma - 1} = \mu_1 + \frac{2c_1}{\gamma - 1}$ (Riemann invariant). In this connection, we have

$$\frac{c_1}{c_o} = \sqrt{\frac{p_1 v_1}{p_o v_o}} = \sqrt{y_i \left(\frac{y_i + \beta}{1 + \beta y_i} \right)}, \quad \frac{c_2}{c_1} = \left(\frac{p_2}{p_1} \right)^{\frac{\gamma - 1}{2\gamma}} = \left(\frac{p_2'}{p_o'} \cdot \frac{p_o}{p_1} \right)^{\frac{\gamma - 1}{2\gamma}}$$

and hence

$$\mu_2 - \mu_1 = \frac{2(c_1 - c_2)}{\gamma - 1} = \frac{2c_o}{\gamma - 1} \sqrt{y_i \left(\frac{y_i + \beta}{1 + \beta y_i} \right)} \left[1 - \left(\frac{y_t}{y_i} \right)^{\frac{\gamma - 1}{2\gamma}} \right] \tag{2.33r}$$

From Fig. 2.5(a), we may write $\mu_1 = \mu_2' - (\mu_2 - \mu_1)$ which gives

$$\frac{y_i - 1}{\sqrt{1+\beta y_i}} = \sqrt{\frac{\rho_o}{\rho_o'}}\left(\frac{\beta'-1}{\beta-1}\right)\frac{y_t - 1}{\sqrt{1+\beta' y_t}} - \frac{2}{\gamma-1}\sqrt{\frac{\gamma y_i}{\beta-1}\left(\frac{y_i+\beta}{1+\beta y_i}\right)}\left[1-\left(\frac{y_t}{y_i}\right)^{\frac{\gamma-1}{2\gamma}}\right] \tag{2.33}$$

by direct substitution of Eqs. (2.32i), (2.32t), and (2.33r) with sonic velocity $c_o = \sqrt{\gamma p_o v_o}$.

Numerical solution of Eq. (2.33) can be sought for y_t and other shock parameters as before.

2.5 Oblique shock reflection and diffraction

When a normal shock wave encounters a rectangular block, a sequence of interactions occur as shown in Fig. 2.6. The incident shock S_i is reflected at the front rigid wall as a reflected shock S_r and the overriding S_i. In addition, a rarefaction wave R sets forth as a result of the pressure difference $p_r > p_i$. The corner flow

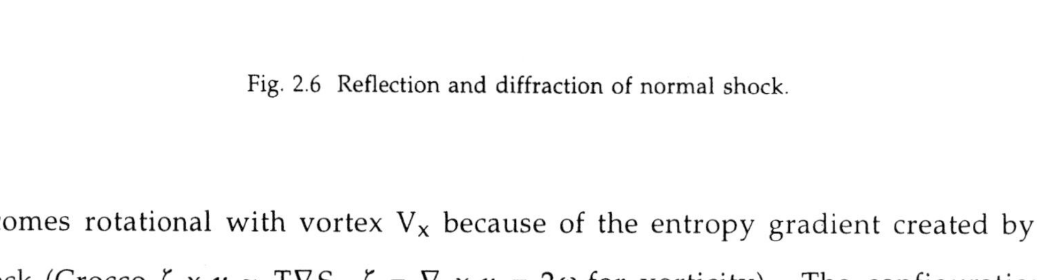

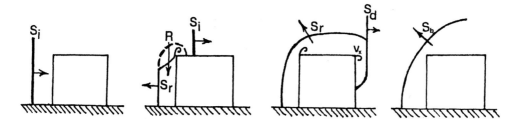

Fig. 2.6 Reflection and diffraction of normal shock.

becomes rotational with vortex V_x because of the entropy gradient created by the shock (Crocco $\zeta \times \boldsymbol{u} \approx T\nabla S$, $\zeta = \nabla \times \boldsymbol{u} = 2\omega$ for vorticity). The configuration of diffracted shock S_d appears to be the result of a Mach reflection for strong shock S_i. If S_d degenerates into sound (weak shock), the bow shock S_b is left over as a detached shock. Thus a one-dimensional shock propagation turns out to exhibit two dimensionality and obliquity during reflection and diffraction.

A second example is depicted interestingly in Fig. 2.7. When a normal shock S_i

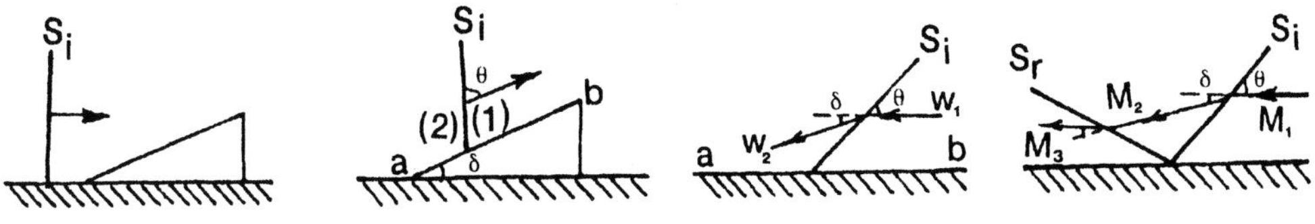

Fig. 2.7 Supersonic RR of oblique shock ($M_1 > M_2 > M_3 > 1$). Note normal shock S_i in laboratory coordinates and oblique shocks in moving coordinates.

goes uphill over a wedge, it is actually an oblique shock if the inclined plane ab is reoriented horizontally (cf. Fig. 2.1). Supersonic streamline flow requires that a reflected shock S_r be generated if the deflection angle δ is smaller than the critical δ_x of Eq. (2.20). This is the supersonic regular reflection (RR), Fig. 2.8(a), to be accounted for later by a two-shock theory. For stronger shock S_i with larger angle θ

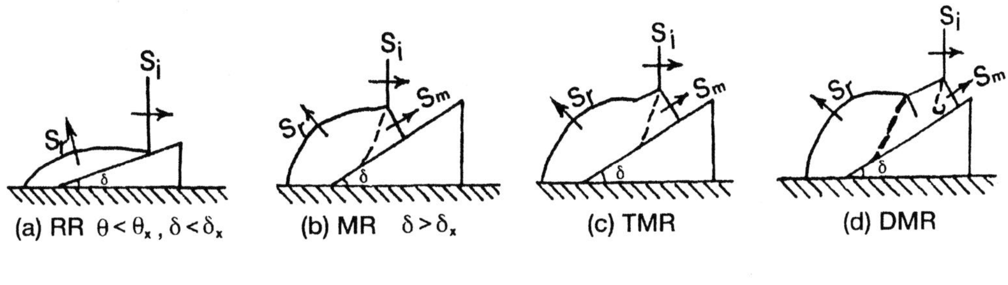

Fig 2.8 Four modes of oblique shock reflection: (a) regular reflection, (b) Mach reflection, (c) transitional Mach reflection, and (d) double Mach reflection.

and $\delta > \delta_X$, the reflection becomes subsonic irregular modes, the simpler one of which is known as Mach reflection (MR), Fig. 2.8(b), to be accounted for by a three-shock theory. Note that these two reflections correspond to the two branches of shock polar in Fig. 2.2. Other modes of oblique shock reflection may be mentioned as Fig. 2.8(c) and (d). Note that all reflected shocks S_r turn around the wedge vertex (or concave corner), featuring the diffraction effect. Thus shock reflections and diffractions are closely associated in one event. When a shock is diffracted around a convex corner, Mach reflection will be accompanied by rarefaction waves (Prandtl-Meyer expansion) and vortices as already noted in the first example.

Before proceeding further, it will be helpful to be conversant with the properties of oblique SW by reviewing Section 2.2. Let us refer to Eqs. (2.17) - (2.21) which are pivotal for the analysis of oblique shock reflection. Yet another formulation of Eq. (2.17) is of interest here. From Fig. 2.1 we may write

$$w_2^2 = u_2^2 + b^2 = u_2^2 + u_1^2 \cot^2\theta = u_1^2\left[\left(\frac{\rho_1}{\rho_2}\right)^2 + \cot^2\theta\right]$$

$$p_2 - p_1 = \rho_1 u_1^2\left(1 - \frac{\rho_1}{\rho_2}\right), \quad \frac{p_2}{p_1} = \frac{(\gamma+1)\rho_2 - (\gamma-1)\rho_1}{(\gamma+1)\rho_1 - (\gamma-1)\rho_2}$$

and hence

$$M_2^2 = \left(\frac{w_2}{c_2}\right)^2 = \frac{\rho_2 u_1^2}{\gamma p_2}\left[\left(\frac{\rho_1}{\rho_2}\right)^2 + \cot^2\theta\right] = \frac{\rho_2}{\rho_1}\left(\frac{p_2 - p_1}{\gamma p_2}\right)\frac{\left(\frac{\rho_1}{\rho_2}\right)^2 + \cot^2\theta}{\left(1 - \frac{\rho_1}{\rho_2}\right)}$$

$$= \frac{2\left[\left(\frac{\rho_1}{\rho_2}\right)^2 + \cot^2\theta\right]}{\frac{\rho_1}{\rho_2}\left[(\gamma+1) - (\gamma-1)\frac{\rho_1}{\rho_2}\right]} \qquad (2.17b)$$

This will prove very useful in the form

$$\cot^2\theta = \frac{M_2^2}{2}\left[(\gamma+1)\frac{\rho_1}{\rho_2} - \left(\gamma - 1 + \frac{2}{M_2^2}\right)\left(\frac{\rho_1}{\rho_2}\right)^2\right] \tag{2.34}$$

which has a minimum at $(\rho_1/\rho_2)_n = (\gamma+1)M_2^2[2(\gamma-1)M_2^2+4]^{-1}$, viz.

$$\cot^2\theta_n = \frac{(\gamma+1)^2 M_2^4}{8[(\gamma-1)M_2^2+2]} \tag{2.35}$$

Let $\theta = \theta_a$ for $M_2 = 1$ in Eqs. (2.34) which results in

$$\cot^2\theta_a = \frac{\gamma+1}{2}\left[\frac{\rho_1}{\rho_2} - \left(\frac{\rho_1}{\rho_2}\right)^2\right] \tag{2.34a}$$

For $\theta_n = \theta_{an}$ and $M_2 = 1$, Eq. (2.35) gives

$$\cot^2\theta_{an} = \frac{\gamma+1}{8} \tag{2.35a}$$

Let $\theta = \theta_b$ for $M_2 = M_b$ which is a constant to be determined by Eqs. (2.17b) and (2.35). Now we have

$$\cot^2\theta_b = \frac{M_b^2}{2}\left[(\gamma+1)\frac{\rho_1}{\rho_2} - \left(\gamma - 1 + \frac{2}{M_b^2}\right)\left(\frac{\rho_1}{\rho_2}\right)^2\right] \tag{2.34b}$$

and

$$\cot^2\theta_{bn} = \frac{(\gamma+1)^2 M_b^4}{8[(\gamma-1)M_b^2+2]} \tag{2.35b}$$

Let us now examine Eqs. (2.34) - (2.35b) graphically as sketched in Fig. 2.9 which appears to provide three zones of relevance. Thus curve $90° - \theta_{an} - \theta_a$ marks the border line of zone (A), and curve $90° - \theta_{bn} - \theta_b$ delineates the bounds of zone (C).

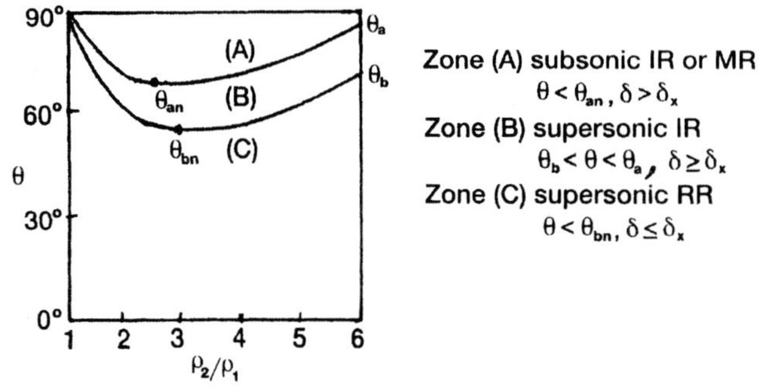

Fig. 2.9 Regular and irregular reflections of oblique SW in three domains.

Zone (B) is the third in between, which is also important. Note the correspondence between curve 90° - θ_{bn} - θ_b of Fig. 2.8 and branch θ_o - θ_w - θ_s of Fig. 2.2 for $M_1 > M_2 \geq 1$. Therefore zone (C) covers the domain of supersonic regular reflection with $\theta < \theta_{bn}$ and $\delta \leq \delta_x$ (note θ_w for weak SW). Referring to Figs. 2.2 and 2.9 again, we notice curve 90° - θ_{an} - θ_a corresponding to the subsonic branch θ_s - θ_v - 90° (now strong shock $\theta_v > \theta_w$ as well as $\theta_a > \theta_b$). Therefore zone (A) belongs to subsonic irregular reflection, viz. Mach reflection with $\theta > \theta_{an}$ and $\delta > \delta_x$. Zone (B) fits the domain of supersonic irregular reflection, viz. shock detachment with $\theta_b < \theta < \theta_a$ and $\delta \gtrsim \delta_x$. For more detail about these results, see Reference 21, pp. 223-248.

As mentioned earlier, the regular reflection of an oblique shock may be analyzed by a two-shock theory for S_i and S_r. Let us generalize Eqs. (2.17b) and (2.20) as

$$M_{j+1}^2 = \frac{(\gamma+1)^2 M_j^4 \sin^2\theta_j - 4\left(M_j^2\sin^2\theta_j - 1\right)\left(\gamma M_j^2 \sin^2\theta_j + 1\right)}{\left[2\gamma M_j^2 \sin^2\theta_j - (\gamma-1)\right]\left[(\gamma-1)M_j^2\sin^2\theta_j + 2\right]} \tag{2.17d}$$

and

$$\cot\delta_j = \sqrt{\frac{\sin^2\theta_j}{1-\sin^2\theta_j}}\left[\frac{(\gamma+1)M_j}{2\left(M_j^2\sin^2\theta - 1\right)} - 1\right] \tag{2.20a}$$

respectively, with j = 1, 2 for computing $S_i(M_1, \theta_1 \rightarrow \delta_1, M_2)$ and $S_r(M_2, \delta_1 \rightarrow \theta_2, M_3)$. Here we have set $\delta_2 = \delta_1$ to facilitate the computation, and the angles of incidence and reflection are given by $\theta_i = \theta_1$ and $\theta_r = \theta_2 - \delta_2$ respectively. For example, given S_i with $M_1 = 2.0$ and $\theta_1 = 35°$ (Fig. 2.10), we obtain $\delta_1 = 6°$ and $M_2 = 1.8$. Because the

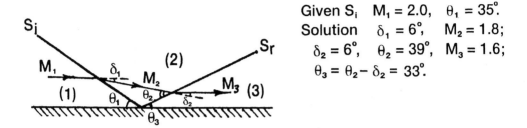

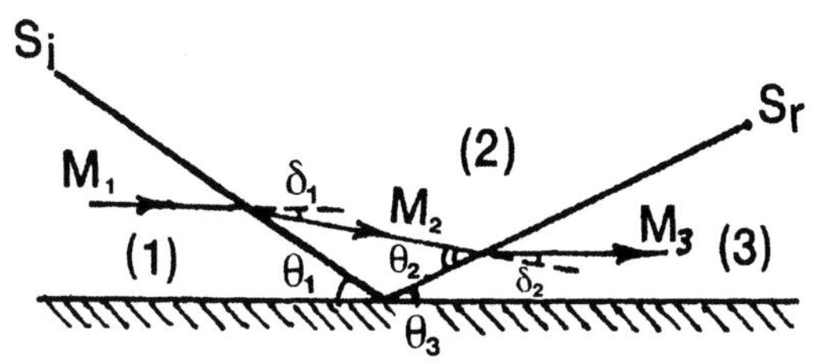

Given S_i $M_1 = 2.0$, $\theta_1 = 35°$.
Solution $\delta_1 = 6°$, $M_2 = 1.8$;
$\delta_2 = 6°$, $\theta_2 = 39°$, $M_3 = 1.6$;
$\theta_3 = \theta_2 - \delta_2 = 33°$.

Fig. 2.10 RR of oblique shock according to 2-shock theory.

deflection of flow implies $\delta_2 = \delta_1 = 6°$, we can find $\theta_2 = 39°$ from Eq. (2.20a) and then $M_3 = 1.6$. Note $\theta_i \neq \theta_r$ versus Snell's law elsewhere with $\theta_i = \theta_r$ and $c_i/\sin\theta_i = c_t/\sin\theta_t$. Also note that S_r is a planar shock without diffraction for a flat reflection surface (versus the curved S_r with diffraction over a wedge).

Now that the two-shock theory of RR is well accepted, the three-shock theory of MR is not so completely established. In fact, much of MR is derived from the results of shock-tube experiment. Analytically speaking, MR is attributable to three shocks (viz. an incident S_i, a reflected S_r, and a Mach stem S_m) plus a contact or tangential discontinuity C_d (also known as slip stream or vortex sheet). As it will become clearer by illustration, MR is a result of transition from RR to IR (see Fig. 2.9). For example, given S_i with $M_1 = 2.0$ and $\theta_1 = 47°$ (Fig. 2.11), we obtain $\delta_1 = 16.3°$ and $M_2 = 1.4$ (with $\delta_x = 9.5°$). Note $\delta_1 = \delta_2 > \delta_x$ here. With $M_2 = 1.4$ and $\delta_2 = 16.3°$, we cannot work out a RR with θ_2 and M_3 as in the first example. So a transition (either δ_s or δ_x) must be tried for solution. Assuming $M_3 = 1$, we have $\delta_s = 9°$ and hence $\theta_2 = 63°$.

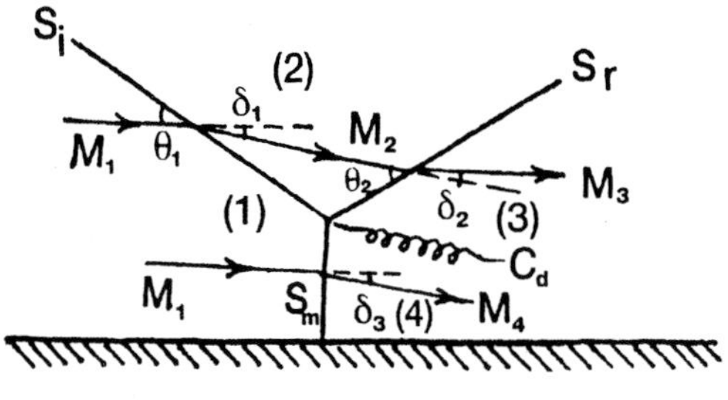

Fig. 2.11 MR of oblique shock according to 3-shock theory.

From Eq. (2.13) we deduce $p_2/p_1 = 2.32$, $p_3/p_2 = 1.65$, and hence $p_3/p_1 = 3.7$ for $\gamma = 1.4$. To account for S_m as a normal shock, we arrive at $M_4 = 0.33$ and $p_4/p_1 = 4.5$. But the 3-shock theory requires $\delta_3 = \delta_1 - \delta_2$ and $p_3 = p_4$ for a complete solution. Our computation indicates $\delta_3 = 7.3°$ and $p_4 > p_3$. Thus a sonic transition with $\delta_2 = \delta_s = 9°$ and $M_3 = 1$ is not acceptable. Let us try a detachment transition with $M_3 = 1.30$, $\delta_2 = \delta_x = 6.54°$. Then we obtain $\theta_2 = 69.4°$, $p_3/p_2 = 1.95$ and hence $p_3/p_1 = 4.53$. These indicate $\delta_3 = 9.8°$ and $p_4 \approx p_3$ as an acceptable solution. Sometimes numerical iterations go nowhere to reach a convergence. Thus no solution is attainable, and that is the shortcome of 3-shock theory. However, other modes of Mach reflection have been identified by shock-tube experiments. For more detail, see Reference 18 (Chapter 7) and Refernce 19.

It should be pointed out that shock diffraction is always present where there are edges, corners, and more complex MR. In this connection, a well-accepted theory is still lacking except for Whitham's GSD (geometrical shock dynamics of Reference 20) cf. GTD (geometrical theory of diffraction) including GO (geometrical optics), GA (geometrical acoustics) and GS (geometrical seismology) which are the ramification of ray theory for wave propagation.

2.6 Summary

It is worthwhile to recapitulate the concepts and theories which are modifiable, generalizable, or analogical on particular occasions. The theory of normal SW is based on gas dynamics which treats the nonlinear, supersonic flow of an inviscid, non-conducting fluid with mathematical discontinuities for its viable solution. Oblique SW are the generalization of normal SW, and they are especially useful for the study of transonic, supersonic, and hypersonic aerodynamics. These SW are planar (one- or two-dimensional), and their weak and strong limits are of heuristic value. Spherical and cylindrical SW are thus treated as strong shocks diverging in simple geometry. The theory of shock reflection, transmission, and diffraction is the result of shock jumps which must satisfy the boundary conditions of given geometry and EOS. The algebraic aspects of these problems are often complex, and so we have taken expeditious steps to obtain solutions. All theories covered so far involve only one EOS (viz. $E = pv/(\gamma - 1)$), and they can serve as a building block for more advanced study of SW.

Referring to Eqs. (2.1b) and (2.5), let us compare a shock jump with an acoustic perturbation as follows:

$$\left(\frac{p_2}{p_1}\right)_H = \frac{\left(\frac{\gamma+1}{\gamma-1}\right)\frac{\rho_2}{\rho_1} - 1}{\frac{\gamma+1}{\gamma-1} - \frac{\rho_2}{\rho_1}} \quad \text{vs.} \quad \left(\frac{p_2}{p_1}\right)_s = \left(\frac{\rho_2}{\rho_1}\right)^{\gamma}$$

$$\left(\frac{U}{c_1}\right)_H^2 = \frac{2\frac{\rho_2}{\rho_1}}{(\gamma+1) - (\gamma-1)\frac{\rho_2}{\rho_1}} \quad \text{vs.} \quad \left(\frac{c_2}{c_1}\right)_s^2 = \left(\frac{\rho_2}{\rho_1}\right)^{\gamma-1}$$

which indicate the weak limit of normal SW to be sonic ($p_2 \approx p_1$, $U = c_1 = c_2$) as $\rho_2 \to \rho_1$. Note that the two EOS exhibit a second-order contact with

$$\frac{\beta+1}{\beta-1}=\left(\frac{dy}{dx}\right)_1=\gamma \quad \text{and} \quad \frac{2(\beta+1)}{(\beta-1)^2}=\left(\frac{d^2y}{dx^2}\right)_1=\gamma(\gamma-1)$$

where $y = p_2/p_1$, $x = \rho_2/\rho_1$, and $\beta = (\gamma+1)/(\gamma-1)$ have been used for simplicity. Elsewhere $d^2y/dx^2 > 0$ is a criterion of shock stability or $s_2 > s_1$ (entropy rise due to the irreversible shock transition). On the other hand, the two shock equations imply the strong limit of $\rho_2/\rho_1 = (\gamma+1)/(\gamma-1)$ as p_2/p_1 and U/c_1 tend to infinity ($M_1 = u_1/c_1 \approx U/c$, U being the shock velocity). Also Eqs. (2.3a) - (2.8a) give all properties of strong SW explicitly. Compare these and Eqs. (2.13a) - (2.17b) with $\theta = 90°$. Oblique SW are weaker than normal SW because of $\theta < 90°$. Eqs. (2.13) - (2.16) indicate the weak limit at $\theta = \arcsin 1/M_1$ (such oblique shock is elsewhere known as Mach wave). Eq. (2.18) implies several important features of the oblique shock theory for $\arcsin 1/M_1 \leq \theta \leq 90°$; there exist a maximum deflection δ_x and a sonic deflection δ_s, which dictate the regular and irregular shock reflections; and the analysis of these shock reflections can be facilitated by using shock polars θ vs. δ (cf. Fig. 2.2), p_2/p_1 vs. δ, and M_2 vs. δ based on Eqs. (2.18), (2.13), and (2.17a).

Chapter 3 Theoretical and Empirical EOS

3.1 The four states of matter

SW take place in all states of matter. As the behavior and properties of these states differ considerably, so does their response to SW. Accordingly it is convenient to divide these states into four which may be described categorically as follows.

(1) Gases are free particles collectively moving at random, macroscopically shapeless (compressible), stable at high T and low p (low ρ).

(2) Liquids may be looked upon as either dense gases (less compressible because of intermolecular forces) or disordered solids lacking tensile and shear strengths; they stay or flow with a free surface under the influence of gravity.

(3) Solids are considered as strongly-interacting lattice-work with elasticity, and they are stable at high p (high ρ) and low T. Other forms such as polymers may break down like solids or flow like liquids; glasses and ceramics exhibit viscoelasticity or flow at glassy temperature; porous materials show anomalous behavior at low ρ.

(4) Plasmas are treated distinctly in three regimes: Cold plasma at ambient T as quasi-neutral, conducting fluid for magnetohydrodynamics; partially ionized gas (e, i^+, a°) at higher T to be modeled by Saha theory; and hot plasma (T > 5,000°K) as fully ionized gas (two-fluid theory) in special environments such as the sun, stars, lightning bolt, nuclear-explosion fireball, etc.

Hundreds of theoretical and empirical EOS are available for the characterization of these states. Yet they are valid in limited ranges. No EOS can be used to admit a change of states or phase transition. The formulation of EOS calls for knowledgeable background of thermodynamics and statistical mechanics, and hence it is pragmatic to seek the proper use of an EOS rather than its derivation. A good EOS must produce data and results verifiable by experiments. Because numerous EOS are used in many fields of study, it is impossible to cover them all in this work. However, we will seek to include a sufficient number of EOS which are theoretically or empirically accurate for the study of SW.

3.2 A theory for the construction of EOS

It is of analytical interest to consider Eq. (1.8), namely,

$$dE = Tds - pdv$$

whence follow

$$p = -\left(\frac{\partial E}{\partial v}\right)_s, \quad T = \left(\frac{\partial E}{\partial s}\right)_v, \quad \left(\frac{\partial p}{\partial s}\right)_v = -\left(\frac{\partial T}{\partial v}\right)_s$$

Thus a thermodynamic formulation of the EOS is $E = E(v,s)$, $p = p(v,s)$, and $T = T(v,s)$. In this connection there are four important parameters as given by

$$\gamma = \frac{c^2}{pv} = -\left(\frac{\partial \ell n p}{\partial \ell n v}\right)_s \tag{3.1}$$

$$\Gamma = \frac{\alpha c^2}{c_p} = \frac{v}{T}\left(\frac{\partial p}{\partial s}\right)_v = -\left(\frac{\partial \ell n T}{\partial \ell n v}\right)_s \tag{3.2}$$

$$g = \frac{pv}{E} = -\left(\frac{\partial \ell n E}{\partial \ell n v}\right)_s \tag{3.3}$$

and

$$j = \frac{Ts}{E} = \left(\frac{\partial \ell n E}{\partial \ell n s}\right)_v \tag{3.4}$$

where the identity $(\partial p/\partial v)_s(\partial v/\partial s)_p(\partial s/\partial p)_v = -1$ has been used in Eq. (3.2) with $\alpha = v^{-1}(\partial v/\partial T)_p$ = thermal expansivity, $c^2 = -v^2(\partial p/\partial v)_s = (\partial p/\partial \rho)_s$ for acoustic velocity c and $c_p = T(\partial s/\partial T)_p$ = specific heat at constant pressure. Note also

$$\Gamma = v\left(\frac{\partial p}{\partial E}\right)_v = \frac{v}{c_v}\left(\frac{\partial p}{\partial T}\right)_v = \frac{v\alpha B_t}{c_v} = \frac{\alpha c^2}{c_p}$$

for the Grüneisen parameter Γ with $B_t = -v(\partial p/\partial v)_T$ = isothermal bulk modulus.

Moreover, we may write

$$\left(\frac{\partial \ell n \mathrm{T}}{\partial \ell n s}\right)_v = \frac{s}{c_v} = j - 1 \tag{3.5}$$

and

$$\left(\frac{\partial \ell n p}{\partial \ell n s}\right)_v = \frac{\mathrm{T}s}{pv}\Gamma = j\frac{\Gamma}{g} \tag{3.6}$$

in compliment to form two groups of EOS parameters. Note that Eq. (3.5) follows from $c_v = (\partial \mathrm{E}/\partial \mathrm{T})v = \partial(j^{-1}\mathrm{T}s)/\partial \mathrm{T} = (s + c_v)/j$, assuming j = constant for specialty to become clear later. Specified properly, the four parameters (γ, Γ, g, j) will prove very useful for the construction of EOS for classical ideal gases, quantum ideal gases, and solids.

While statistical mechanics provides an elegant formulation of EOS, its mathematical representation is much more obscure and complicated. Generally speaking, Gibbs canonical distribution is the backbone of statistical mechanics. The microcanonical and grand canonical (or macrocanonical) distributions may be regarded as degenerate canonical distributions. Gibbs grand canonical distribution can also lead to Bose-Einstein and Fermi-Dirac distributions, which may again degenerate into Boltzmann distribution. Note that each of these distributions has its limits to model identical or different particles, with or without interactions. It is of fundamental importance to introduce Gibbs canonical distribution as follows.

The probability function of Gibbs distribution is $w_i = \mathrm{A}\bar{e}^{\beta\varepsilon_i}$ with ε_i denoting the ith particle's energy, A and β the Lagrange multipliers to be determined below. According to statistics, we have $\sum_i w_i = 1$ and $\mathrm{E} = <\varepsilon_i> = \sum_i \varepsilon_i w_i$ for normalizing and averaging procedures, respectively. Thus we may write

$$1 = \sum_i w_i = \mathrm{A}\sum_i \bar{e}^{\beta\varepsilon_i} = \mathrm{AZ},$$

where $\mathrm{Z} = \sum_i \bar{e}^{\beta\varepsilon_i} = \frac{1}{\mathrm{A}}$ is the partition function. From Boltzmann law for the entropy, we have

$$s = -k < \ell n w_i >= -k\sum_i \ell n\left(A\bar{e}^{\beta\varepsilon_i}\right)w_i$$

$$= -k\ell n A\sum_i w_i + k\beta\sum_i \varepsilon_i w_i = -k\ell n A + k\beta E$$

Differentiating this once, we obtain

$$\frac{1}{T} = \left(\frac{\partial s}{\partial E}\right)_v = k\beta$$

and

$$s = k\ell n Z + (E/T) \tag{3.7a}$$

This is essentially

$$F = E - Ts = -kTlnZ \tag{3.7b}$$

for the Helmholtz free energy F, k being the Boltzmann constant. From Eqs. (3.7b) and (1.8), we deduce

$$dF = -pdv - sdT \tag{3.7c}$$

$$p = -\left(\frac{\partial F}{\partial v}\right)_T = kT\left(\frac{\partial \ell n Z}{\partial v}\right)_T \tag{3.8}$$

$$s = -\left(\frac{\partial F}{\partial T}\right)_v = k\frac{\partial}{\partial T}(T\ell n Z)_v \tag{3.9}$$

and

$$E = F + Ts = kT^2(\partial \ell n Z/\partial T)_v \tag{3.10}$$

These provide a statistical formulation of EOS as F = F(v,T), p = $p(v, T)$, s = $s(v, T)$, and E = E(v, T), all to be determined by a given partition function $Z = \sum_i \bar{e}^{\frac{\varepsilon_i}{kT}}$. Note also that Eq. (3.10) is directly deducible from

$$E = \sum_i \varepsilon_i w_i = \sum_i \frac{\varepsilon_i \bar{e}^{\beta\varepsilon_i}}{\sum \bar{e}^{\beta\varepsilon_i}} = -\frac{\partial}{\partial\beta}\left[\ell n\left(\sum_i \bar{e}^{\beta\varepsilon_i}\right)\right]$$

$$= -k\partial(\ell nZ)\big/\partial\left(\tfrac{1}{T}\right) = kT^2\left(\frac{\partial \ell nZ}{\partial T}\right)_v$$

The foregoing discussions offer only a glimpse at the theory of EOS. In what follows, we will supplement this with the various EOS applicable to the study of SW.

3.3 Classical ideal gases

The EOS, pv = RT, for an ideal gas is a classical expression of simplicity and applicability. It is an empirical result of thermodynamics, and now it can be derived from three different approaches (viz. analytical thermodynamics, gas kinetics, and statistical mechanics). Much of the important development has relied on the use of this simple EOS for the theory of aeronautics, astronautics, and astrophysics (sun, stars). Of course, it is an especially useful tool for the theory of SW.

Let us refer to the total mass M in volume V and with N particles (m = molecular weight with molar volume V_o and N_o particles). Thus we have mass density $\rho = M/V = m/V_o$ ($v = 1/\rho$ = specific volume), number density $n = N/V = N_o/V_o$, and $mR = R_o = N_o k$ (R = specific gas constant). Note the three quantities: $R_o = 8.314 \times 10^7$ erg/°Kmole = universal gas constant, $k = 1.38 \times 10^{-16}$ erg/°K = Boltzmann constant, and $N_o = 6.023 \times 10^{23}$ particles/mole, all at standard T and p. Now the ideal-gas EOS may be written as

$$pV = MRT = M\left(\frac{N_o k}{m}\right)T = NkT \tag{3.11a}$$

$$pv = RT \text{ for } M = 1 \text{ or } v = V/M \tag{3.11b}$$

$$p = \rho RT = n\left(\frac{m}{N_o}\right)RT = nkT \tag{3.11c}$$

$$pV_o = N_o kT = R_o T \text{ for 1 mole } (m) \tag{3.11d}$$

which are complementary to one another. Also they are widely used in many fields of study. It is worth noting that the partition function of Boltzmann distribution is

$$Z = \frac{z^N}{N!} = \frac{1}{N!}\left[\left(\frac{mkT}{2\pi\hbar^2}\right)^{\frac{3}{2}} V\right]^N$$

Substituting this in Eq. (3.8), we obtain

$$p = NkT\frac{\partial}{\partial V}\left\{\ell n\left[\left(\frac{mkT}{2\pi\hbar^2}\right)^{\frac{3}{2}}\frac{eV}{N}\right]\right\}_T = \frac{NkT}{V}$$

which is the statistical mechanics derivation of Eq. (3.11a), with e = base of natural *ln* and $\hbar = (6.626 \times 10^{-27})/2\pi$ erg.sec = Planck constant.

Eq. (3.11c) is suitable for the environment of low ρ and high T. In astrophysics the gas is a mixture of completely ionized hydrogen, helium, and others with the following composition (mass fraction $x + y + w = 1$):

	n^+	n^-
H	$\frac{x\rho}{m_H}$	$\frac{x\rho}{m_H}$
He	$\frac{y\rho}{4m_H}$	$\frac{y\rho}{2m_H}$
others	$\frac{w\rho}{\alpha m_H}$	$\frac{\zeta w\rho}{\alpha m_H}$

$${}^1_1H_0 \Leftrightarrow H^+ + e^-$$

$${}^4_2He_2 \Leftrightarrow He^{++} + 2e^-$$

$${}^\alpha_\zeta A_\nu \quad \alpha = \zeta + \nu \quad (\zeta \text{ protons, } \zeta \text{ electrons})$$

$$\frac{1+\zeta}{\alpha} \approx \frac{\zeta}{\alpha} \approx \frac{1}{2}$$

The number density may be written as

$$n = \sum_i n_i = \left(2x + \frac{3}{4}y + \frac{1}{2}w\right)\frac{\rho}{m_H}$$

and total pressure is

$$p = \sum_i p_i = \sum_i n_i kT = \rho \frac{kT}{m_H}\left(2x + \frac{3}{4}y + \frac{1}{2}w\right)$$

$$= \frac{\rho kT}{m_H}\left(\frac{1}{2} + \frac{3}{2}x + \frac{1}{4}y\right) = \frac{\rho kT}{\mu m_H} \tag{3.11e}$$

with $\frac{1}{\mu} = \frac{1}{2} + \frac{3}{2}x + \frac{1}{4}y$ and μm_H = mean atomic mass. For partially ionized hydrogen ($H^0 \Leftrightarrow H^+ + e^-$, $n^+ = n^-$, and $m_H = m^+ >> m_e$), the degree of ionization is defined as $\delta = n^+/(n^0 + n^+)$. Noting $\rho = n^0 m_H + n^+(m^+ + m_e) \approx (n^0 + n^+)m_H$, we may write

$$n^- = n^+ = (n^0 + n^+)\delta = \delta\rho/m_H\,, \quad n^0 = (1-\delta)\rho/m_H$$

$$p = \sum_i n_i kT = (1+\delta)\rho kT/m_H \tag{3.11f}$$

$$\frac{n^+ n^-}{n^0} = \left(\frac{\delta^2}{1-\delta}\right)\frac{\rho}{m_H}$$

For ionization equilibrium, we must have

$$\frac{n^+ n^-}{n^0} = \left(\frac{m_e kT}{2\pi\hbar^2}\right)^{\frac{3}{2}} \exp\left(-\frac{I}{kT}\right)$$

where I is the ionization potential. Combining the last two expressions, we obtain the Saha equation

$$\frac{\delta^2}{1-\delta} = \frac{m_H}{\rho}\left(\frac{m_e kT}{2\pi\hbar^2}\right)^{\frac{3}{2}} \exp\left(-\frac{I}{kT}\right) \tag{3.12}$$

for partially ionized hydrogen at low ρ and high T ($0 < \delta < 1$). However, Eq. (3.11f) tends to the same limit of complete ionization $\delta = 1$ ($n^+ >> n^0$) as Eq. (3.11e) with $y = 0$ and $x = 1$.

Thermodynamically speaking, the classical ideal gas calls for R, c_v, and c_p to be all constant (so that $R = c_p - c_v$) in the EOS, pv = RT and E = c_vT. The results of letting $\gamma = c_p/c_v$ may be summarized as follows:

$$E = c_v T = \frac{RT}{\gamma - 1} = \frac{pv}{\gamma - 1}$$

$$\Gamma = g = \gamma - 1 \text{ being all constant}$$

$$T_s v^{\gamma-1} = \text{constant}, \quad p_s v^{\gamma} = \text{constant}$$

where Eqs. (3.1) - (3.3) are used with amendments of

$$\gamma = \frac{c_p}{c_v} = \frac{h}{E} = \frac{dh}{dE} = -\left(\frac{\partial \ell np}{\partial \ell nv}\right)_s$$

and

$$\Gamma = \frac{v}{T}\left(\frac{\partial p}{\partial s}\right)_v = v\left(\frac{\partial p}{\partial E}\right)_v = \frac{pv}{E}$$

These properties of the classical ideal gas and the following notes suffice for us to denominate it as the γ-law gas when a particular constant is specified to its γ. From the theory of gas kinetics and strong shock limit, we have

$$\gamma = \frac{c_p}{c_v} = 1 + \frac{2}{f} \qquad (f = \text{ number of freedom degrees})$$

$$\gamma_{i,\ell} = \frac{c_p}{c_v} = \frac{6i-3}{6i-5} \qquad (i \text{ atoms in a line})$$

$$\gamma_{i,n} = \frac{c_p}{c_v} = \frac{3i-2}{3i-3} \qquad (i \text{ atoms not in line})$$

$$\beta = \frac{\gamma+1}{\gamma-1} = 1 + f \qquad (\frac{\rho_2}{\rho_1} = \beta \text{ as } \frac{p_2}{p_1} >> 1, \ \frac{T_2}{T_1} >> 1)$$

Note that γ is less predictable by these equations if the molecular composition has $i >$ 3 atoms. Let *td* = translational degrees of freedom, *rd* = rotational degrees, and *vd* = vibrational degrees. Under normal conditions it is plausible to consider $i = 1, f = 3$ for 3*td*; $i = 2, f = 5$ for (3*td* + 2*rd*); and $i = 3, f = 6$ for (3*td* + 3*rd*) for gases such as A, He; H_2, N_2, O_2; and CO_2, H_2O respectively. Thus we have

(a)

i =	1	2	3
f =	3	5	6
γ =	1.67	1.40	1.33
β =	4	6	7

(b)

i =	2	3
f =	7	13 (?)
$\gamma_{i,\ell}$ =	1.29	1.15
β =	8	14 (?)

(c)

$i = 3$	$i = 3$
$f = 13$	$f = 7$
$\gamma_{3,\ell} = 1.15$	$\gamma_{3,n} = 1.17$
$\beta = 14$	$\beta = 8$

(d) specified estimate for air

$f = 19$

$\gamma = 1.10$

$\beta = 20$

It should be remarked that β increases as γ decreases to 1.10 extremely low. In (a) through (c) an upper bound of $f \sim 12$ is acceptable in view of *td, rd,* and *vd* accountable only. In (d) the estimate is based on $T_2 \sim 10^6$°K of the very strong shock due to air burst of nuclear explosion [$T_2/T_1 = 2\gamma(\gamma-1)M_1^2/(\gamma+1)^2 \sim 10^4$ for $M_1 \sim 500$ and $\gamma = 1.10$. See Eq. (2.6a)]. Note now $f = \beta - 1 = 19$ for 3*td* + 3*rd* + 6*vd* + 7 other degrees such as dissociation, ionization, electron excitation, which are too complex to count. Such a γ-law gas description is oversimplified because c_p, c_v, and R are no longer constant at high T.

Before passing, it is worth mentioning that the γ-law gas is also useful to model the detonation products (dense gases) of high explosive with $\gamma = \dfrac{\rho_o D^2}{p_j} - 1 \approx 3$ and $\rho_j/\rho_o = (\gamma+1)/\gamma$ (see section 4.3).

3.4 Quantum ideal gases

At high ρ, at high T or very low ρ, or at high p, gases and condensed matter become degenerate (dissociated, ionized, quantized, relativistic). Such a state may be modeled by the quantum ideal gas versus the classical ideal gas. For pragmatic purposes we will seek to deal with only a special class of such gases with constant indices (g, j) in Eqs. (3.3) and (3.4). As a result we have $\Gamma = g = \gamma - 1$ again (with $p_s v^{g+1} = p_s v^{\gamma}$ = constant and $Tv^g = Tv^{\gamma-1}$ = constant), but c_v and c_p are no longer constant. From Eq (3.5) we have $c_v = s/(j\text{-}1)$ and $c_p = (\partial h/\partial T)_p = (g + 1)s/[j - (g + 1)]$ with $h = E + pv = (g + 1)Ts/j$. Note $\gamma = g + 1$, $c_p/c_v = (g + 1)(j - 1)/[j - (g + 1)]$ = constant, but $\gamma \neq c_p/c_v = (j - 1)\gamma/(j - \gamma)$. While the adiabatic index γ is defined by Eq. (3.1), the expression

$$c_p/c_v = B_s/B_t = 1 + \Gamma\alpha T \tag{3.13}$$

will prove useful to distinguish the two parameters accurately. For a proof of the identity (3.13), see the Reference 5, pp. 295, 297, and 334. Here we have used the standard symbolism: $c_v = (\partial E/\partial T)_v = T(\partial s/\partial T)_v$, $c_p = (\partial h/\partial T)_p = T(\partial s/\partial T)_p$, $B_t = -v(\partial p/\partial v)_T$, $B_s = -v(\partial p/\partial v)_s = c^2/v$, $\Gamma = v(\partial p/\partial E)_v = \alpha v B_t/c_v = \alpha c^2/c_p$, $\alpha = v^{-1}(\partial v/\partial T)_p$. These are essentially the thermodynamic properties which can serve to test or validate a given EOS expediently.

For this inquiry it is suitable to summarize the equilibrium relations among the 5 state variables (p, v, T, s, E), 6 energy functions (pv, Ts, E, h, F, G), with h denoting enthalpy, F free energy, and G free enthalpy, and 9 properties $(c_v, c_p, B_t, B_s, \alpha, \Gamma, \gamma, g, j)$ as given by Eqs. (3.14) - (3.18f):

$$pv = gE, \quad jE = Ts \tag{3.14}$$

$$h = E + pv = (1 + g)E \tag{3.15}$$

$$F = E - Ts = (1 - j)E \tag{3.16}$$

$$G = E + pv - Ts = (1 + g - j)E \tag{3.17}$$

$$dE = Tds - pdv = E\left(\frac{j}{s}ds - \frac{g}{v}dv\right) \tag{3.14a}$$

$$dh = Tds + vdp = \frac{h}{1+g}d\left(\ell n s^j p^g\right) \tag{3.15a}$$

$$dF = -sdT - pdv = -\frac{F}{1-j}d\left(\ell n T^{j} v^{g}\right) \tag{3.16a}$$

$$dG = -sdT + vdp = \frac{G}{1+g-j}d\left(\ell n T^{-j} p^{g}\right) \tag{3.17a}$$

$$\Gamma = g = \gamma - 1 = \text{ constant} \tag{3.18a}$$

$$c_v = \frac{s}{j-1} = \left(\frac{j}{j-1}\right)\frac{E}{T} \neq \text{ constant} \tag{3.18b}$$

$$c_p = \frac{(1+g)s}{j-(1+g)} = \frac{j}{j-(1+g)}\frac{h}{T} \neq \text{ constant} \tag{3.18c}$$

$$\frac{c_p}{c_v} = \frac{B_s}{B_t} = 1 + \Gamma\alpha T = \frac{(1+g)(j-1)}{j-(1+g)} = \text{ constant} \tag{3.18d}$$

$$\frac{B_s}{p} = \gamma = 1+g, \quad \frac{B_t}{p} = \frac{j-(1+g)}{j-1} = \text{ constaant} \tag{3.18e}$$

$$\alpha T = j/(j-1-g) = \text{ constant} \tag{3.18f}$$

which provide an explicit formulation of the EOS and properties of the quantum ideal gas with the EOS based on (g, j) = constant. It will become clear that this model includes the classical ideal gas as a special case of $j \to \infty$ and $c_p/c_v = 1 + g = \gamma$.

A glance at Eqs. (3.14a) - (3.17a) reveals that they can provide four schemes to deduce the alternative EOS explicitly. For illustration, two of these schemes are given by integrating Eqs. (3.14a) and (3.16a) as follows:

$$E(v,s) = c_1 v^{-g} s^{j}, \quad p(v,s) = g c_1 v^{-1-g} s^{j}, \text{ and}$$

$$T(v,s) = j c_1 v^{-g} s^{j-1} \tag{3.14b}$$

$$F(v,T) = c_2 v^{\frac{g}{j-1}} T^{\frac{j}{j-1}}, \quad p(v,T) = -\left(\frac{\partial F}{\partial v}\right)_T = \frac{g c_2}{1-j} v^{\frac{1+g-j}{j-1}} T^{\frac{j}{j-1}},$$

$$s(v,T) = -\left(\frac{\partial F}{\partial T}\right)_v = \frac{j c_2}{1-j} v^{\frac{g}{j-1}} T^{\frac{1}{j-1}} \text{and } E(v,T) = \frac{c_2}{1-j} v^{\frac{g}{j-1}} T^{\frac{j}{j-1}} \tag{3.16b}$$

It is interesting to note that indices (g, j) determine the properties of the quantum ideal gas explicitly as given by Eqs. (3.18a) - (3.18f), c_v and c_p being not constant.

Specific values of (g, j) can serve to identify particular environments of matter and hence to determine its EOS and physical properties.

First of all, we must have $g > 0$ and $j \geq 0$ as the lower limit according to Eq. (3.14). Clasusius Virial theorem may be written as

$$3pv = 2E_k + E_p \quad (E_k = \text{kinetic energy, } E_p = \text{potential energy})$$

for ideal gases. In a non-relativistic and non-interacting environment, the gas has $E_k = E$, $E_p = 0$, and hence $g = 2/3$. In the extreme environment with relativistic effect but no interaction, we must have $E_k = E/2$ (note $E = mc^2$, $E_k = mc^2/2 = E/2$ for photons) and $E_p = 0$. Thus we infer the upper limit $g = pv/E = (2E_k + E_p)/3 = 1/3$. On the other hand, quantum ideal gases include the classical ideal gas as a special case with $j \to \infty$ and $c_p/c_v = 1 + g = \gamma$. Accordingly the two indices must fall in the ranges:

$0 < g < 2/3$

$0 \leq j \leq \infty$

Note that there are three particular cases of interest, namely $j = 0$ (Ts = 0), $j = 1$ (F = 0), and $j = 1 + g$ (G = 0) from Eqs. (3.14), (3.16) and (3.17) respectively. These values of j turn out to shed some light on the EOS of numerous degenerate matters. But the case of $j = 1$ has to be excluded because F = 0 (viz. $E = Ts > 0$, $T \neq 0$, $s \neq 0$) leads to $p = -(\partial F/\partial v)_T = 0$ and $s = -(\partial F/\partial T)_v = 0$. Moreover, Eq. (3.18d) gives $c_p/c_v = 0$ for $j = 1$ (again unacceptable). In general we should have $c_p/c_v \geq 1$. Thus $c_p/c_v < 0$ is also not acceptable for $j < 1$. Below we will consider $j = 0$ and $j > 1$.

A. $g = 1/3, j = 1 + g = 4/3$. (1) photon gas at high T, relativistic bosons $g = 1/3$ with entropy $s = 4aT^3v = 4E/3T$. (2) completely ionized, rarefied gas at high T (or hot plasma with radiation pressure $p_r \sim T^4$), Note $p_r > p_g = \rho RT$ for the ions as classical ideal gas at low ρ. (3) phonons (or quantized oscillations) in Debye solid at high p and low T with $c_p \approx c_v$.

B. $g = 2/3, j = 0$. (1) Completely degenerate electron gas with T = 0 and $s \neq 0$; $\alpha T = 0$, $c_p = c_v \neq 0$; $\gamma = 1 + g = 5/3$ fermions with Fermi energy $F_c(v) = E_c(v) \propto v^{-2/3}$. (2)

superfluid ^{4}He-II with s = 0, T$\neq$ 0 (or Landau rotons, i.e. quantized vortex); non-relativistic bosons, $c_p = c_v$ = 0; quantum acoustics for the thermal waves as the second sound. (3) Thomas-Fermi EOS for electrons to be treated as a degenerate Fermi gas at T = 0°K and nuclei as ideal gas, $pv = 2E/3 = (2e^2/15a_0)\,(2Z)^{7/3}\,(3\pi)^{-2/3}x_0^{1/2}\,\psi_0^{5/2}$. See Eqs. (3.38a, b) and (b5) with $\mu = a_0(9\pi^2/128Z)^{1/3}$.

C. $g = 2/3, j = 1 + g = 5/3$. Superfluid ^{3}He-B to be treated as fermions , again the second sound with ρ_s. Note the two fluids ^{4}He-I and ^{3}He-A as ρ_n with the first sound (ρ_n = normal fluid, ρ_s = superfluid).

D. $g = 2/3, j = 2 > 1 + g$. $G = (1 + g - j)E < 0$ conduction electrons in metals as free electron gas at low T and high p, $F = -E = -Cv^{2/3}T^2$, $p_e = -(\partial F/\partial v) = 2/3Cv^{-1/3}T^2$, $p_\ell = nkT$ (lattice vibrations), $p_e >> p_\ell$; $c_v = s$, $c_p/c_v = 5$, $\alpha T = 6$.

E. $g = 2/f, j \to \infty$. Classical ideal gases at high T and low ρ, $p = \rho RT$ with f = 3, 5, 6, $c_p - c_v = R$, $\alpha T = 1$, $c_p/c_v = 1 + g = \gamma = 1 + 2/f$.

It should be noted that the EOS and physical properties are all explicitly determined with (g, j) = constant in the above examples. See Eqs. (3.14b) and (3.18a) - (3.18f) as well as (3.17b). Both classical and quantum ideal gases are γ-law gases [see Eq. (3.18a)], and so their shock behavior and properties may be readily assessed with the equations of Chapter 2. Nevertheless statistical mechanics of quantum ideal gases is much more involved than our simplified examples.

3.5 Van der Waals fluids

Van der Waals introduced two semi-empirical constants (a, b) in the ideal gas EOS which became

$$(p + a/v^2)(v - b) = RT \tag{3.19}$$

This is a versatile EOS for fluids (gas, vapor, liquid), but it is only qualitatively usable because of its unreal characterization of the two-phase isotherm. Perhaps that is the reason for at least 60 modified forms to be worth noting (e.g. Dieterici,

Bethelot, Redlich-Kwong EOS, etc.).

Eq. (3.19) defines a critical state (p_c, v_c, T_c) based on $(\partial p/\partial v)_T = (\partial^2 p/\partial v^2)_T = 0$, which determines $a = 3p_c v_c^2$, $b = v_c/3$, and $p_c v_c/RT_c = 3/8$. Note also $b = 4nvv_m$ with $v_m = \pi d^3/6$ = a single molecule's volume (d = its diameter, n = number density of particles). Thus constant a provides the correction for cohesion or surface tension effects, and constant b for the co-volume of gas molecules. Let us consider $(\partial E/\partial T)_v = c_v$ = constant. In view of $(\partial E/\partial v)_T = T(\partial p/\partial T)_v - p = a/v^2$, we obtain the caloric EOS

$$E = c_v T - \frac{a}{v} = \left(\frac{v-b}{\lambda-1}\right)\left(p+\frac{a}{v^2}\right) - \frac{a}{v} \tag{3.20}$$

with $\lambda = 1 + R/c_v$ = constant. From Eqs. (3.19) and (3.20) follow

$$\alpha = \frac{1}{v}\left(\frac{\partial v}{\partial T}\right)_p = \frac{v-b}{Tv}\left[1-\frac{2a(v-b)^2}{RTv^3}\right]^{-1} \tag{3.19a}$$

$$\Gamma = \frac{v}{c_v}\left(\frac{\partial p}{\partial T}\right)_v = \frac{R}{c_v}\left(\frac{v}{v-b}\right) = v\left(\frac{\lambda-1}{v-b}\right) \tag{3.19b}$$

$$c_p = \left(\frac{\partial h}{\partial T}\right)_p = c_v + \left(p+\frac{a}{v^2}\right)\left(\frac{\partial v}{\partial T}\right)_p = c_v + R\left[1-\frac{2a(v-b)^2}{RTv^3}\right]^{-1} \tag{3.20a}$$

$$\frac{c_p}{c_v} = 1 + \frac{R}{c_v}\left[1-\frac{2a(v-b)^2}{RTv^3}\right]^{-1} \tag{3.20b}$$

$$s = c_v \ell n\left[T(v-b)^{\lambda-1}\right] = c_v \ell n\left[R^{-1}(p+a/v^2)(v-b)^{\lambda}\right] \tag{3.21}$$

$$T_s(v-b)^{\lambda-1} = \exp(s_o/c_v) = T_o(v_o-b)^{\lambda-1} = \text{constant A} \tag{3.21a}$$

$$(p_s + a/v^2)(v-b)^{\lambda} = RA \tag{3.21b}$$

$$\gamma = -\left(\frac{\partial \ell n p}{\partial \ell n v}\right)_s = \frac{\lambda v}{v-b} + \frac{a}{p_s v^2}\left(\frac{\lambda v}{v-b} - 2\right) \tag{3.21c}$$

$$c^2 = -v^2 dp_s / dv = \frac{\lambda v^2}{v-b}\left(p_s + \frac{a}{v^2}\right) - \frac{2a}{v} = \lambda R T_s \left(\frac{v}{v-b}\right)^2 - \frac{2a}{v} \tag{3.21d}$$

When the parameter a is zero, the above equations reduce to the form for Abel's EOS which is used in interior ballistics (burnt gases of gun powder). With $a = b = 0$ the foregoing equations become exactly the same as those of the classical ideal gas. These remarks indicate the analytical merit of van der Waals EOS, which still deserves more attention as to be taken up in Section 4.2.

Using Eqs. (3.20) and (1.7), we immediately deduce the shock EOS as

$$\frac{p_2}{p_1} = \frac{(\lambda+1)v_1 - (\lambda-1)v_2 - 2b}{(\lambda+1)v_2 - (\lambda-1)v_1 - 2b} + \frac{2a}{p_1 v_1^2}\left(\frac{v_1}{v_2} - 1\right)\left[\frac{(\lambda-2)v_1 + b(1 + v_1/v_2)}{(\lambda+1)v_2 - (\lambda-1)v_1 - 2b}\right] \tag{3.22}$$

$$\frac{T_2}{T_1} = \left(\frac{p_2}{p_1} + \frac{a}{p_1 v_2^2}\right)\left(1 + \frac{a}{p_1 v_1^2}\right)^{-1}\left(\frac{v_2 - b}{v_1 - b}\right) \tag{3.19c}$$

$$\frac{s_2 - s_1}{c_v} = \ell n\left[\frac{T_2}{T_1}\left(\frac{v_2 - b}{v_1 - b}\right)^{\lambda-1}\right] = \ell n\left[\left(\frac{p_2}{p_1} + \frac{a}{p_1 v_2^2}\right)\left(1 + \frac{a}{p_1 v_1^2}\right)^{-1}\left(\frac{v_2 - b}{v_1 - b}\right)^{\lambda}\right] \tag{3.21e}$$

which again imply the two special cases, viz., Abel ($a = 0$) and ideal gas ($a = b = 0$ for $\gamma = \lambda = 1 + R/c_v = c_p/c_v$ and $\Gamma = R/c_v = \gamma - 1$ all to be constant). Note that the strong shock limit (as $p_2 \to \infty$) is now given by

$$\frac{\rho_2}{\rho_1} = \frac{v_1}{v_2} = \frac{\lambda + 1}{(\lambda - 1) + 2b/v_1} \tag{3.22a}$$

versus Eq. (2.5a) to account for the co-volume effect. It is interesting here to examine a numerical example with the initial state of shock compression at $T_1 = T_c$, $v_1 = 30v_c$, and $p_1 = 0.085p_c$ according to Eq. (3.19) with $T_c = 8a/27bR$, $v_c = 3b$, and $p_c =$

$a/27b^2$ arbitrarily. For $v_1/v_2 = 6$ and $\lambda = 4/3$ (i.e. $c_v = 3R$ arbitrarily), Eqs. (3.22) and (3.19c) give $p_2 = 58.8p_1 = 4.98p_c$ and $T_2 = 8.93T_c$ respectively. This is a tremendous shock! Note $v_1/v_2 = 6.56$ as $p_2 \rightarrow \infty$ according to Eq. (3.22a), cf. $v_1/v_2 = 7$ for ideal gas of $\gamma = 4/3$.

3.6 Tait EOS for liquids

While van der Waals EOS has the analytical merit for examining the shock behavior and properties of fluids (gas, vapor, liquid), Tait EOS can serve as a simple, expedient tool for the inquiry of liquids. The original Tait equation

$$\frac{v_o - v}{v_o p} = \frac{A}{\Pi + p} \tag{3.23}$$

is meant to fit experimental data in a hyperbola, with parameters A and Π to be determined by data fitting. But investigators either misquote or modify the above expression for better representation of liquids to agree with experimental results. Thus Tammanan introduces the isothermal version as

$$-\frac{1}{v_o}\left(\frac{\partial v}{\partial p}\right)_T = \frac{C}{B + p} \tag{3.23a}$$

and hence

$$v(T,p) = v(T,0)\{1 - C\ell n[1 + p/B(T)]\} \tag{3.23b}$$

Once an empirical function is established for $B = B(T)$, the above EOS determines the pvT-relation readily. Yet another version is the Murnaghan isentrope

$$-v\left(\frac{\partial p}{\partial v}\right)_s = B_o + B_o^{'} p \tag{3.23c}$$

and hence

$$p_s = \frac{B_o}{B_o'}\left[\left(\frac{v_o}{v}\right)^{B_o'} - 1\right] \tag{3.23d}$$

with constants $B_o = -(vdp_s/dv)_o = \rho_o c_o^2$ and $B_o' = d(-vdp_s/dv)/dp_s = -1-(vd^2p_s/dv^2)_o(dp/dv)_o^{-1}$. This EOS is also applicable to solids.

The well-known Bridgman isotherm, with much of the high-pressure data sources up to the early 1940's, is simply expressed as

$$(v_o - v)/v_o p_t = a - bp_t \tag{3.23e}$$

where the two important constants are $a = B_o^{-1}$ and $b = (B_o' + 1)/2B_o^2$ (or $B_o' = (2b/a^2) - 1$). Apparently Eq. (3.23e) is yet another variant of the Tait EOS, but it is a parabolic fit of experimental data. Recently the EOS of water has been elaborated as

$$v_o p/(v_o - v) = B_o + B_1(T)p + B_2(T)p^2 \tag{3.23f}$$

with sophisticated expressions for $B_1(T)$ and $B_2(T)$. A still more sophisticated EOS is available for sea water (namely, IES 80 = International Equations of State 1980). It should be noted that all these EOS are valid within limited ranges as follows: Eq. (3.23b) T = 20° to 650°C, p = 1 to 2,500 atm or 2.5 kb; Eq. (3.23e) T = 25° to 175°C, p = 5 to 35 kb; Eq. (3.23f) T = 0° to 100°C, p = 1 to 1000 atm because of its acoustic data sources; and IES more or less similar to Eq. (3.23f). The Murnaghan Eq. (3.23d) is applicable to similar Tp-ranges (note $p_s(v) > p_t(v)$ and $T_s(v)$ depending upon (v_0, T_0).

Since 1950 shock compression of liquids has boosted the pressure range considerably. Thus Rice and Walsh propose a new EOS based on the assumption

$$(\partial h/\partial v)_p = \zeta(p) = \exp(A + p/B)$$

with empirical constants A = 5.02 and B = 143.15 kb for water. Their experimental results fall within the range of T = 175° to 1000°C and p = 25 to 250 kb. Analytical consideration could extend the upper bound to 450 kb, while the latest experimental results have been reported to be 1.5 Mb by hypervelocity impact and Mach reflection of strong shock. All these developments seem to be less relevant to the Tait EOS, but we seek to conclude this inquiry with a shock EOS which is closely attributable to the Tait EOS for water.

Let us consider the EOS to be

$$E = v(p + B_o)/(B_o' - 1) \tag{3.24}$$

Substituting this into Eq. (1.8) gives

$$Tds = dE + pdv = (B_o' - 1)^{-1}[(B_o + B_o' p)dv + vdp]$$

and hence

$$-v(\partial p / \partial v)_s = B_o + B_o' p$$

which is the same as Eq. (3.23c). Again substituting Eq. (3.24) into Eq. (1.7), we obtain the shock EOS

$$\frac{\rho_2}{\rho_i} = \frac{v_1}{v_2} = \frac{(B_o'+1)p_2 + (B_o'-1)p_1 + 2B_o}{(B_o'-1)p_2 + (B_o'+1)p_1 + 2B_o} \tag{3.24a}$$

with the strong shock limit $\rho_2/\rho_1 = (B_o'+1)/(B_o'-1)$ as $p_2 \to \infty$. Now combining the equations of mass and momentum conservation with Eq. (3.24a), we deduce

$$\frac{u}{U} = 1 - \frac{\rho_1}{\rho_2} = \frac{2(p_2 - p_1)}{(B_o'+1)(p_2 - p_1) + 2(B_o - p_1)} = \frac{2\rho_1 u U}{(B_o'+1)\rho_1 u U + 2(B_o - p_1)}$$

which is simply the shock Hugoniot

$$U^2 - \tfrac{1}{2}(B_o'+1)uU - c_o^2 = 0 \tag{3.24b}$$

with $\rho_1 = \rho_0$ and $p_1 = p_0 << B_0 = \rho_0 c_0^2$ in the denominator term $2(B_0 - p_1)$. Eq. (3.24b) may also be expanded as

$$U = c_o\left[1 + \left(\frac{B_o'+1}{4c_o}\right)u + \frac{1}{2}\left(\frac{B_o'+1}{4c_o}u\right)^2 - \frac{1}{8}\left(\frac{B_o'+1}{4c_o}u\right)^4 + \ldots\right]$$

For a linearized approximation, $U = c_0 + su$, we obtain $B_0' = 4s - 1$ and $B_0 = \rho_0 c_0^2$. Thus shock Hugoniots and the computed (B_0, B_0') for pure water are comparable as follows

(a) $U = 1.493 + 1.962u - 0.075u^2$, $p_H \sim 15$ kb. $B_0 = \rho_0 c_0^2 = 22.4$ kb $B_0' = 4s - 1 = 6.85$ approximately.

(b) $U = 1.483 + 1.70u$, $p_H \sim 100$ kb (Russian data). $B_0 = 21.9$ kb, $B_0' = 5.80$.

(c) $U = 1.80 + 1.60u$, $p_H \sim 150$ kb (Los Alamos data). $B_0 = 32.1$ kb, $B_0' = 5.40$.

(d) Elsewhere isentropic data to fit Eq. (3.23d) give $B_0 = 21.8$ kb and $B_0' = 7.47$.

Note that the discrepancies are due to shock strength and nonlinear shock Hugoniot. The linear shock Hugoniots are taken to exclude phase transitions.

3.7 Grüneisen EOS for solids

Since the 1940's the study of EOS for solids has been enriched remarkably by the probe of shock waves. As a result many books and papers appear to interpret the shock behavior and properties of solids. Here we will seek to provide a concise coverage of the subject as follows: (i) the original Mie-Grüneisen EOS for ideal crystalline solids, (ii) the Debye theory of Grüneisen Γ, (iii) the Los Alamos formulation and its linear shock Hugoniot $U = c_0 + su$, and (iv) our semi-analytical approach to shock EOS.

(i) Mie consider the interaction potential of two atoms to be describable by $\varphi = -a/l^m + b/l^n$ with l = their separation distance and a, b, m, and n all empirical constants (the repulsive $n > m$ cohesive exponent). From this expression Grüneisen deduces

$$E_c = \frac{c_1}{v^{n/3}} - \frac{c_2}{v^{m/3}} = \frac{E_0}{m-n}\left\{ m\left(\frac{v_{00}}{v}\right)^{n/3} - n\left(\frac{v_{00}}{v}\right)^{m/3} \right\} \qquad (3.25a)$$

$$E_0 = -9v_{00}B_0/mn$$

$$p_c = -\frac{dE_c}{dv} = \frac{3B_o}{n-m}\left\{\left(\frac{v_{oo}}{v}\right)^{\frac{n}{3}+1} - \left(\frac{v_{oo}}{v}\right)^{\frac{m}{3}+1}\right\} \tag{3.25b}$$

$$B_o' = (dB_c/dp_c)_o = (m+n+6)/3 \approx 5 \text{ for ideal solids}$$

$$\Gamma_o = (m+n+3)/6 = (B_o' - 1)/2 \approx 2$$

with $B_c = -vp_c' = -vdp_c/dv =$ bulk modulus and subscript o referring to $v = v_{oo}$ at $p_c = 0$ and T = 0°K. The Grüneisen constant Γ_0 is derived from relations involving either thermal expansion or lattice vibration. Note that constants c_1 and c_2 are to be determined by stable crystal structure with $v = Nr^3$, N being the specific number of lattice points and r the length of cube side. Also it is plausible to note that Eq.(3.25b) has several special analogs such as the EOS of Born (n = 9, m = 1), of Birch (n = 4, m = 2), and of Murnaghan ($n = 3B_0' - 3$, m = -3). By postulation of the thermal agitation of lattice points like the gas behavior of $p_{th}v/E_{th} = g = \Gamma_0$ [see Eq. (3.3)], the original Mie-Grüneisen EOS was given as

$$pv + G(v) = \Gamma_o E \tag{3.26a}$$

$$G(v) = \frac{3v_o B_o}{m-n}\left\{\left(\frac{v_o}{v}\right)^{n/3} - \left(\frac{v_o}{v}\right)^{m/3}\right\} = -p_c v$$

with $E = E_c + E_{th}$ = internal energy and $p = p_c + p_{th}$. Apparently a more general formalism of the EOS should be

$$p - p_c = \frac{\Gamma}{v}(E - E_c) \tag{3.26b}$$

for an arbitrary E_c with $p_c = -dE_c/dv$, admitting the macroscopic $\Gamma = v(\partial p/\partial E)_v = -(d\ell n T/\partial \ell n v)_s$ to be volume - dependent only.

(ii) In order to evaluate the specific heat of solids at low T, Debye uses the statistics of phonon (quantized vibration) to establish

$$E_{th} = 3RTD\left(\frac{\theta}{T}\right) \quad \text{with } R = Nk \tag{3.27a}$$

$$D(x) = \frac{3}{x^3}\int_0^x \frac{y^3 dy}{e^y - 1} = \text{Debye function,} \quad x = \frac{\theta}{T}$$

$$F = E - Ts = E_c + 3RT\int \frac{D(x)}{x} dx$$

$$p = -\left(\frac{dF}{dv}\right)_T = -\frac{dE_c}{dv} - 3RT\left(\frac{d}{dx}\int \frac{D(x)}{x} dx\right)\frac{dx}{dv}$$

$$= p_c - 3RTD(x)\frac{d\ell nx}{dv} = p_c - \left(\frac{d\ell n\theta}{dv}\right)E_{th} \tag{3.27b}$$

with $\theta = \hbar\omega / k$ = Debye temperature depending on v only ($x = \theta / T$ and ω = circular frequency of vibration). Now the microscopic Grüneisen parameter of Debye is

$$\Gamma = -d\ell n\theta / d\ell nv = -d\ell n\omega / d\ell nv \tag{3.27c}$$

for the validation of Eq. (3.26b) by Eq. (3.27b). Since the frequency of lattice vibration $\omega / 2\pi = \nu = c / \lambda \propto v^{1/6} B_c^{1/2}$, Eq. (3.27c) leads to the expression

$$\Gamma = -\frac{1}{6} - \frac{1}{2}\frac{d\ell nB_c}{d\ell nv} = \frac{1}{2}\left(B_c' - \frac{1}{3}\right) = -\frac{v}{2}\left(\frac{p_c''}{p_c'}\right) - \frac{2}{3} \tag{3.28a}$$

with $p_c' = dp_c / dv$, $B_c' = dB_c / dp_c$, and $p_c'' = d^2 p_c / dv^2$. This very expression induces considerable controversy, and it is elsewhere generalized as

$$\Gamma = \left(\frac{t}{2} - \frac{2}{3}\right) - \frac{v}{2}\left(\frac{(p_c v^t)''}{(p_c v^t)'}\right) = \frac{B_c}{2}\left(\frac{B_c' - t}{B_c - tp_c}\right) - \frac{1}{6} \tag{3.28b}$$

with $t = 0$ for Slater, $t = 2/3$ for Dugdale-MacDonald, and $t = 4/3$ for Vashchenko-Zubarev, all from different approaches. Eq. (3.28b) yields

$$\Gamma_o = \frac{1}{2}(B_o' - t) - \frac{1}{6} \tag{3.28c}$$

and

$$\Gamma_\infty = \frac{1}{2}B_\infty' - \frac{1}{6} \tag{3.28d}$$

at $p_c = 0$ and very high pressure ($B_c >> p_c$) respectively. Note that the Dugdale-MacDonald value Γ_0 agrees with the Grüneisen constant favorably. All three versions of Eq. (3.28b) reduce to Eq. (3.28d) in favor of the Slater Eq. (3.28a). It is worth noting that Eq. (3.28d) implies $B' = dB/dp = B/p = c^2/pv = \gamma =$ constant. Thus we infer $\Gamma_\infty = 2/3$ for $B_\infty' = \gamma = 5/3$ (monatomic ideal gas). See Eqs. (3.1) and (3.18a). Now the typical Grüneisen data $B_0' = 5$ and $\Gamma_0 = 2$ plausibly lead to $B_\infty' = 5/3$ and $\Gamma_\infty = 2/3$ by a factor of one third respectively.

(iii) Shock-wave research at Los Alamos National Laboratory has generated a rich source of EOS data and results. In favor of the Dugdale-MacDonald formula, the following formulation was first explored numerically with experimental results of $p_H(v)$ as input:

$$-\frac{v}{2}\left[\frac{\left(p_c v^{2/3}\right)''}{\left(p_c v^{2/3}\right)'}\right] - \frac{1}{3} = \Gamma = \frac{v(p_H - p_c)}{\frac{1}{2}(v_o - v)p_H + \int_{v_{oo}}^{v} p_c dv} \tag{3.29}$$

Note that v_0 is the initial value of v at $p_H = 0$ and $T_0 \neq 0$. In view of $v_o^{-1}(\partial v/\partial T)_p = \alpha = \Gamma_o c_p/c_o^2$, we have

$$v_{oo} = v_o\left(1 - \Gamma_o c_p T_o / c_o^2\right) \tag{3.29v_{oo}}$$

for computational purposes. Let us consider

$$ds = \left(\frac{\partial s}{\partial T}\right)_v dT + \left(\frac{\partial s}{\partial v}\right)_T dv = \left(\frac{\partial s}{\partial T}\right)_v \left[dT + \left(\frac{\partial T}{\partial s}\right)_v \left(\frac{\partial s}{\partial v}\right)_T dv\right]$$

$$= \frac{c_v}{T}\left[dT - \left(\frac{\partial T}{\partial v}\right)_s dv\right] = \frac{c_v}{T}\left(dT + \frac{\Gamma}{v} T dv\right)$$

where Eq. (3.2) and $(\partial T/\partial s)_v(\partial s/\partial v)_T(\partial v/\partial T)_s = -1$ have been used. Applying the above expression with Eq. (1.8) to the Rankine-Hugoniot, we obtain

$$c_v\left(dT_H + \frac{\Gamma}{v}T_H dv\right) = Tds = \frac{1}{2}(v_o - v)dp_H + \frac{1}{2}p_H dv$$

and hence

$$\frac{dT_H}{dv} + \frac{\Gamma}{v}T_H = \frac{1}{2c_v}\left[(v_o - v)\frac{dp_H}{dv} + p_H\right] \tag{3.30a}$$

For constant c_v and $b = \Gamma_0/v_0 = \Gamma/v$, Eq. (3.30a) can be integrated to give

$$T_H = T_o + \frac{1}{2c_v}\left[(v_o - v)p_H + e^{-bv}\int_{v_o}^{v} e^{bx}\left[2 - b(v_o - x)\right]p_H dx\right] \tag{3.30b}$$

for the assessment of shock heating. It may be noted that Eq. (1.8) implies

$p_c = -dE_c/dv$ for the isotherm at $T_o = 0°K$

or

$p_s = -dE_s/dv$ for the isentrope at $ds = 0$

These curves appear to be parallel to each other except for a shift from v_{oo} at $p_c = 0$ and $T_c = 0°K$ to v_0 at $p_s = 0$ and $T_0 = 273°K$. See Eq. (3.29v_{00}) for the difference between v_{oo} and v_0. In this connection, Eq. (3.26b) may be exploited to deduce

$$\frac{dE_s}{dv} + \frac{\Gamma}{v}E_s = \left[\frac{\Gamma}{2v}(v_o - v) - 1\right]p_H(v) \tag{3.31a}$$

which is integrable to give

$$E_s = E_o + e^{-bv}\int_{v_o}^{v} e^{bx}\left[\frac{b}{2}(v_o - x) - 1\right]p_H(x)dx \tag{3.31b}$$

and

$$p_s = -\frac{dE_s}{dv} = \left[1 - \frac{b}{2}(v_o - v)\right] p_H(v) + be^{-bv} \int e^{bx} \left[\frac{b}{2}(v_o - x) - 1\right] p_H(x) dx \tag{3.31c}$$

for $b = \Gamma / v$ as before. Perhaps the most significant result at Los Alamos is the linear shock Hugoniot and EOS $p_H(v)$

$$U = c_o + su = c_o / (1 - s\varepsilon) \tag{3.32a}$$

$$\varepsilon = u / U = 1 - v / v_o$$

$$p_H = \rho_o u U = \rho_o c_o^2 \varepsilon / (1 - s\varepsilon)^2 \tag{3.32b}$$

for many solids (metals, non-metals, plastics, etc.). These empirical EOS are well-received worldwide today. Note $p_s(v_o) = p_H(v_o) = 0$, $p_s'(v_o) = p_H'(v_o) = -c_o^2/v_o^2$, $p_s''(v_o) = p_H''(v_o) = 4sc_o^2/v_o^3$, $p_H'''(v_o) = -18s^2c_o^2/v_o^4$, $p_s'''(v_o) = p_H'''(v_o) + \Gamma_o p_H''(v_o)/2v_o = 2sc_o^2 v_o^{-4}(\Gamma_o - 9s)$, $B_o = -v_o p_s'(v_o) = \rho_o c_o^2$, $B_o' = -1 - v_o p_H''(v_o) / p_H'(v_o) = 4s - 1$, $-B_o B_o'' = (v p_H''/p_H')_o^2 - (v p_H'''/p_H')_o - (v^2/p_H')_o (p_H''' + \Gamma p_H''/2v)_o = 2s(\Gamma_o + 2 - s)$ being all useful properties of Eq. (3.32b).

(iv) From the atomic-binding point of view, all crystalline solids fall in one or another among the following classes:

Solids	Lattice points	Binding forces	Examples
I. molecular crystals	molecules	van der Waals cohesion	N_2, H_2O, CO_2
II. ionic crystals	ions	Coulomb attraction	NaCl, KNO_3
III. covalent crystals	atoms sharing electrons	electron - pairing	C, SiC, SiO_2
IV. metals	nuclei	lattice vibration	Ag, Cu, Fe

Accordingly we may take a semi-analytical approach to the study of shock EOS by using the lattice potentials as follows.

I. molecular $E_C = Ae^{-br} - Cr^{-6}$

II. ionic $E_C = Ae^{-br} - Cr^{-1}$

III. covalent $E_c = (Ar^{-1} - C)e^{-br}$

IV. metallic $E_c = Ae^{-br} - Cr^{-1} + Dr^{-2}$

with $r \propto v^{1/3}$ these potentials are analogous to Eq. (3.25a), but the repulsive part is modified from a quantum statistical approximation. Note that the empirical constants A, b, C, D are to be determined from thermodynamic data E_0, B_0, B_0', Γ_0 at $p_c = 0$, $v = v_{oo}$, and T = 0°K. For metals at pressures not too high, the Fermi-energy term Dr^{-2} may be neglected. Then the intrinsic difference from ionic E_c lies in the fact that metals have conduction electrons like a free electron gas intermingling with nuclei. Now from Eq. (3.26b) we may compute the shock EOS:

$$p_H = \left(\frac{dE_c}{dv} + \frac{\Gamma}{v}E_c\right) \Big/ \left\{\frac{\Gamma}{2v}(v_o - v) - 1\right\} \tag{3.33a}$$

$$U = v_o\sqrt{p_H/(v_o - v)}$$

$$u = \sqrt{(v_o - v)p_H}$$

with E_c appropriately chosen from the above list. This is an inverse manipulation to examine the shock data which may be known or unknown for a given solid. Of course the computation is simple with $\Gamma = \Gamma_0 = \alpha c_0^2/c_p$ or $b = \Gamma/v = \Gamma_0/v_0$. However, it should improve the result further to use $\Gamma/\Gamma_0 = (v/v_0)^\kappa$ with the exponent κ deducible from Eqs. (3.28b) and (3.28c) as follows.

$$\frac{d\Gamma}{d\ell nv} = -B\frac{d}{dp}\left[\frac{B}{2}\left(\frac{B'-t}{B-tp}\right)\right] = -\frac{B}{2}\left\{B'\left(\frac{B'-t}{B-tp}\right) + \frac{BB''}{B-tp} - B\left(\frac{B'-t}{B-tp}\right)^2\right\}$$

$$\kappa = \left(\frac{d\ell n\Gamma}{d\ell nv}\right)_o = \frac{-B_oB_o'' - t(B_o' - t)}{B_o' - t - ⅓} \tag{3.33b}$$

Since the available compression data are accurate at most to the second order only, we seek to compute $-B_0B_0''$ from a generalized compressibility equation of Tait or

Murnaghan:

$$B_s/B_o = (P/P_o)^M \quad \text{with } p_s = P - P_o \tag{3.34a}$$

$$B'_s/B'_o = (v/v_o)^N \tag{3.34b}$$

where M, N and P_o are all positive constants to be determined. From these follow

$$M = B_o'^2/(B_o'^2 - B_oB_o'') = B_o'/(B_o' + N) \tag{3.34c}$$

$$N = -B_oB_o''/B_o' = (1-M)\,B_o'/M \tag{3.34d}$$

$$P_o = MB_o/B_o' = B_o/(B_o' + N) \tag{3.34e}$$

$$p_s = P_o\left\{\left(\frac{B_s}{B_o}\right)^{1/M} - 1\right\} = \frac{B_o}{B_o' + N}\left\{\left(1 + N\ell n\frac{v_o}{v}\right)^{\frac{B_o'}{N}+1} - 1\right\} \tag{3.34f}$$

Since Eqs. (3.34a) and (3.34b) yield the Murnaghan EOS exactly with M = 1 and N = 0, it turns out that the above five equations as a whole require

$$\frac{1}{2} < M < 1$$

$$B_o' > N > 0$$

$$B_o'^2 > -B_oB_o' > 0$$

for the determination of κ as well [see Eq. (3.33b)]. Let us first use $B_o' = 4s - 1$ and $-B_oB_o'' = 2s(\Gamma_o + 2 - s)$ which are given earlier as a result from the linear shock Hugoniot. For $B_o' = 5$, we obtain

$-B_oB_o''$	M	N	κ	*t*
8.5	0.75	1.7	1.82	0
7.5	0.77	1.5	1.15	2/3
6.5	0.79	1.3	0.48	4/3

Now let us compare the results from Eqs. (3.34c), (3.34d), and (3.33b) only:

M	N	$-B_oB_o''$	κ	t
0.77	1.5	7.5	1.15	2/3
0.80	1.25	6.25	1.33	0
0.83	1.0	5.0	1.07	0

With $\Gamma/v \approx \Gamma_o/v_0$ and $B_o' \approx 5$ as a guide, we infer $-B_oB_o'' > B_o'(N > 1)$ and hence $\kappa > 1$. Also Eq. (3.34f) can serve as a third-order isentrope with $M < 1$.

3.8 EOS for plastics, mixtures, powders, and porous material

Pure substances exhibit shock behavior and properties which are attributable to their EOS as discussed in the foregoing sections. It should be remarked that the simple shock Hugoniots Eqs. (2.8), (3.24b), and (3.32a) have become very influential for the study of SW and EOS. Thus Eq. (3.32a) can serve as a gauge to fit SW data of many solids including mixtures (e.g. alloys, rocks) and plastics (e.g. polystyrene, polyvinyl chloride, polymethyl methacrylate, polytetrafluoroethylene). These plastics are polymers of large molecules with chemical formulas and trade names as follows:

PS (polystyrene), $(C_8H_8)_n$ styrene
PVC (polyvinyl chloride), $(C_2ClF_3)_n$ vinyl
PMMA (polymethyl methacrylate), $(C_5H_8O_2)_n$ Plexiglas
PTFE (polytetrafluoroethylene), $(C_2F_4)_n$ teflon

Their constitutive equations are elsewhere the main theme of viscoelasticity, but their shock EOS are simply fitted with experimental results at high pressure as listed below.

	$\rho_o[g/cm^3]$	$c_o[mm/\mu sec]$	s	Γ_o
PS	1.04	2.75	1.32	
PMMA	1.18	2.57	1.54	0.97
Teflon	2.15	1.68	1.82	0.59

Note that the Grüneisen constant is based on $\Gamma_o = \dfrac{\alpha c_o^2}{c_p}$. According to a two - dimensional theory the Grüneisen parameter of long-chain polymers is given by

$$\Gamma = \frac{B'}{2} - \frac{1}{4}$$
$$B_o' = (m+n+4)/2 \approx 8$$
$$\Gamma_o = (m+n+4)/4 \approx B_o'/2 = 4$$

which must, however, be used discriminatorily. Using the above EOS data, we can deduce isentropes, isotherms, and other properties of those polymers.

A theory of mixture is made simple by formulating its Gibbs free enthalpy, $G(p,T) = \sum_j x_j G_j(p_j, T_j) = \sum_j x_j G_j(p,T)$, with component mass fractions $\sum x_j = \sum_j m_j/m = 1$ (j = 1, 2 ..). Thus we have

$$E + pv - Ts = G = \sum_j x_j E_j + p\sum_j x_j v_j - T\sum_j x_j s_j$$

$$E = \sum_j x_j E_j \text{ or } u^2 = \sum_j x_j u_j^2 \text{ with } u_j = \frac{c_{jo}}{2s_j}\left\{\left(1 + \frac{4s_j p}{\rho_{jo} c_{jo}^2}\right)^{\frac{1}{2}} - 1\right\}$$

$$v = \sum_j x_j v_j = \sum_j x_j v_{jo}\left(c_{jo} + s_j u_j\right)^{-1}\left[c_{jo} + \left(s_j - 1\right)u_j\right]$$

$$s = \sum_j x_j s_j \text{ also verifiable by } s = -(\partial G/\partial T)_p = -\sum_j x_j\left(\partial G_j/\partial T_j\right)_{p_j}$$

$U = u(1 - v/v_o)^{-1}$ to be checked with $p = uU/v_o$

where the component shock Hugoniots $U = c_{jo} + s_j u_j$ have been used. From these we can plot the data points (u, U) to see whether or not they fall in a straight line to fit Eqs. (3.32a) and (3.32b) usefully. If needed, the Grüneisen constant $\Gamma_o = \alpha v B_t / c_v$ of a mixture can be calculated by using

$$\alpha v = (\partial v/\partial T)_p = \sum_j x_j \alpha_j v_j$$

$$B_t = -v(\partial p/\partial v)_T = v\left[-\sum x_j (\partial v_j/\partial p_j)_{T_j}\right]^{-1} = v\left(\sum_j x_j v_j / B_j\right)^{-1}$$

$$c_p = T(\partial s/\partial T)_p = \sum_j x_j c_{pj} \approx c_v$$

which is verifiable with $\Gamma = v\left(\dfrac{\partial p}{\partial E}\right)_v = v\left(\sum x_j v_j / \Gamma_j\right)^{-1}$. All these calculations for a mixture are based on the known EOS and properties of its components.

It is now of interest to examine porous and granular materials which have some common grounds regarding their properties and EOS. Here by porous material we mean a solid with density reduced by distention (due to void or air inclusion). Plastic foam, sintered metal, and pressed explosive are familiar examples. Granular materials are aggregates such as grain, sand, snow, powder, etc. Note that porous materials share the chemical bond and properties of their original solids, which may be weaker or modified by the change of densities. But the behavior and properties of granular materials are more dominated by the stratification effect of gravity. Being of different nature though, the porosity of these materials turns out to be an important parameter for our inquiry. Since porosity ϕ is the volume fraction of voids ($\phi = V_h/V$), the ratio of solid density ρ_0 to porous material density ρ_0^+ is known as distensibility $\mu_o = \rho_0/\rho_0^+ = [M/(V - V_h)](M^+/V)^{-1} \approx (1 - \phi)^{-1}$. Note $M \approx M^+$, $0 < \phi < 1$, and $\mu_o > 1$. We will use μ_o as a measure of porosity.

It is of analytical interest to consider the snow-plow model for shock compression of porous material as shown in Fig. 3.1. The voids are crushed without

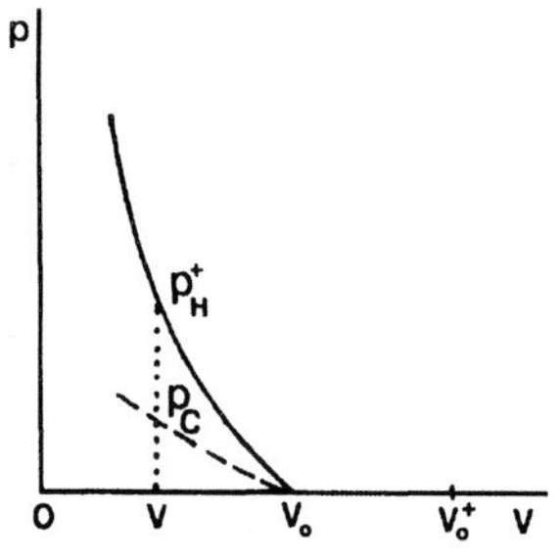

Fig. 3.1 Snow - plow model for shock compression of porous material with zero resistance ($p = 0$) along $v_o^+ v_o$.

resistance (zero-pressure line $v_0{}^+v_0$, v_0 being the solid state), and the Hugoniot curve is depicted as a whole by $v_0{}^+v_0 p_H{}^+$. Let curve $v_0 p_c$ denote the cold compression of solid at T = 0°K. Assuming Eq. (3.26b) to be applicable to porous material with $\Gamma = \Gamma_0$ (Grüneisen constant of solid), we may write

$$p_H^+ = \frac{\dfrac{2}{\Gamma_o} p_c - \dfrac{2E_c}{v}}{\left(\dfrac{2}{\Gamma_o} + 1\right) - \dfrac{v_o^+}{v}} \tag{3.35a}$$

for the porous Hugoniot curve $v_0{}^+v_0 p_H{}^+$ of Fig. 3.1. Such a shock EOS may exhibit four kinds of behavior as shown in Fig. 3.2, namely

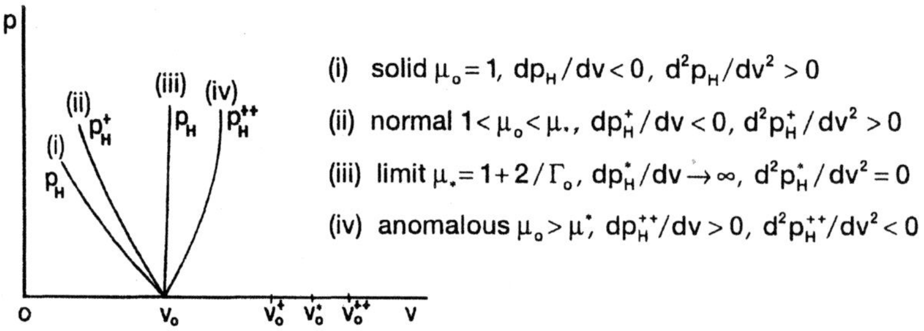

Fig 3.2 Shock behavior of porous material (snow-plow).

(i) solid Hugoniot $v_0 p_H$ ($v_0^+ = v_0$), $\mu_0 = 1$, $dp_H/dv < 0$, $d^2p_H/dv^2 > 0$, shock stability ($p_2 > p_1$, $v_2 < v_1$, $s_2 > s_1$).

(ii) porous Hugoniot $v_0^+ v_0 p_H^+$, $1 < \mu_0 < \mu_*$, $dp_H^+/dv < 0$, $d^2p_H^+/dv^2 > 0$, stable shock compression ($p_2 > p_1$, $v_2 < v_1$, $s_2 > s_1$).

(iii) limit Hugoniot $v_0^* v_0 p_H^*$, $\mu^* = 1 + 2/\Gamma_0$, $dp_H^*/dv \to \infty$, $d^2p_H^*/dv^2 = 0$, strong shock limit $p_H^+ \to \infty$ in Eq. (3.35a) or $p_2 \to \infty$, $v_2 = v_1 = v_0$ ($s_2 > s_1$).

(iv) abnormal Hugoniot $v_0^{++} v_0 p_H^{++}$, $\mu_0 > \mu_*$, $dp_H^{++}/dv > 0$, $d^2p_H^{++}/dv^2 < 0$, thermodynamic instability ($p_2 > p_1$, $v_2 > v_1$) or anomalous behavior due to shock heating and melting ($p_2 > p_1$, $v_2 > v_1$, $s_2 > s_1$).

For porous material at low distention, we may substitute p_c and neglect $E_c << p_c$ in Eq. (3.35a) with

$$p_c = -\int_{v_o}^{v} B\frac{dv}{v_o} = \frac{B_o}{v_o}(v_o - v) \approx \frac{B_o}{\rho_o}(\rho - \rho_o)$$

$$E_c = -\int_{v_o}^{v} p_c dv \approx \frac{B_o}{2\rho_o^3}(\rho - \rho_o)^2 << p_c$$

Accordingly the shock EOS, Eq. (3.35a), becomes

$$p_H^+ = \frac{B_o\left(\frac{\rho}{\rho_o} - 1\right)}{\left(1 + \frac{\Gamma_o}{2}\right) - \left(\frac{\mu_o \Gamma_o}{2}\right)\frac{\rho}{\rho_o}} \qquad (3.35b)$$

which can be transformed into the shock Hugoniot

$$U^2 - \left(1 + \frac{\Gamma_o}{2}\right)uU - c_o^2\left[\mu_o + (1 - \mu_o)\frac{U}{u}\right] = 0 \qquad (3.35c)$$

Eq. (3.35b) has proved in good agreement with experimental results at high pressures for metals of low porosity, and its strong shock limit gives $\rho/\rho_o^+ = \mu_* = 1 + 2/\Gamma_o = 4$ (monatomic $\Gamma_o = \gamma - 1 = 2/3$). Note that Eq. (3.35c) with $\mu_o = 1$ reduces to the form of Eqs. (2.8) and (3.24b). The effect of porosity seems to cause the shock Hugoniot being nonlinear versus Eq. (3.32a). Also note that the snow-plow model turns out to predict poor results at low pressures because of the compressive strength. Despite several other models competing for improvements, the following formulation appears to be more plausible. In contrast to the snow-plow model of Fig. 3.1, the porous Hugoniot is to be determined realistically by curve $v_0{}^+p_H{}^+v^+$ in Fig. 3.3 where curve v_0p_Hv denotes the solid Hugoniot at the same pressure ($p_H = p_H{}^+$ but $v < v^+$). The corresponding states of cold compression are (v^+, $p_c{}^+$) and (v,

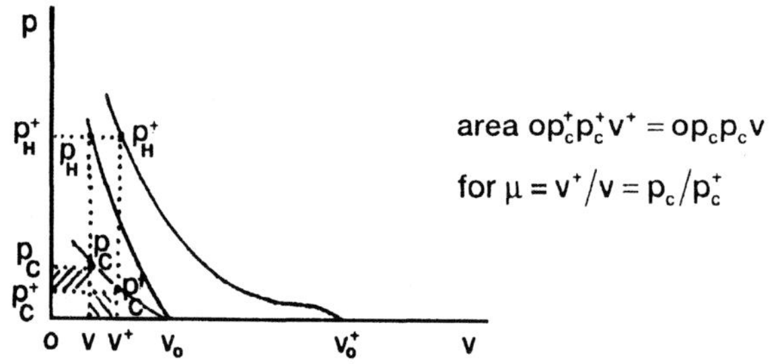

Fig. 3.3 Shock compaction of porous material with elastic-plastic strength taken into account.

p_c) respectively, and they are interlinked by assuming the distention parameter

$$\mu = \rho/\rho^+ = v^+/v = p_c/p_c^+ \tag{3.36a}$$

Apparently Eq. (3.36a) gives the initial value $\mu = \rho_0/\rho_0{}^+ = \mu_0$ ($p_H = p_c = 0$, $v = v_0$; $p_H{}^+ = p_c{}^+ = 0$, $v = v_0{}^+$). Now let us introduce

$$E_H^+ - E_c^+ = E_{th}^+ = \frac{v^+}{\Gamma_o}\left(\mu p_{th}^+\right) = \frac{\mu v^+}{\Gamma_o}\left(p_H^+ - p_c^+\right)$$

for Mie-Grünesien correlation. Substituting the Rankine-Hugoniot equation, we deduce

$$p_{\mathrm{H}}^{+} = \frac{\dfrac{2\mu}{\Gamma_o} p_c^{+} - \dfrac{2E_c^{+}}{v^{+}}}{\left(\dfrac{2\mu}{\Gamma_o} + 1\right) - \dfrac{v_o^{+}}{v^{+}}} \tag{3.36b}$$

for the shock EOS (curve $v_o{}^{+}p_{\mathrm{H}}{}^{+}v^{+}$ of Fig. 3.3). This is a step further to improve the snow-plow model of Eq. (3.35a). It is of relevance to refer to the isotherm of powder compaction

$$\frac{\rho}{\rho_o^{+}} = \frac{1 + bp}{1 + abp}$$

with $1/a = \rho_o/\rho_o^{+} = \mu_o$ ($\rho = \rho_o$ as $p \to \infty$) and $b = 1/B_o(1-a) = \mu_o/B_o(\mu_o - 1)$ from $B_o/\rho_o^{+} = dp/d\rho$ at $p = 0$. This is to be modified as

$$\frac{\rho^{+}}{\rho_o^{+}} = \frac{1 + bp_c^{+}}{1 + ab\left(\dfrac{B_o - p_c^{+}}{B_o}\right)p_c^{+}} = f(p_c^{+}) \tag{3.37a}$$

for the cold compression of porous material. From this and Eq. (3.36a) we may write

$$\mu\rho_o^{+} f(p_c^{+}) = \mu\rho^{+} = \rho = \frac{\rho_o B_o}{B_o - p_c} = \frac{\rho_o B_o}{B_o - \mu p_c^{+}}$$

and hence

$$\mu^2 - \frac{B_o \mu}{p_c^{+}} + \frac{\mu_o B_o}{p_c^{+} f(p_c^{+})} = 0 \tag{3.37b}$$

As an earlier expression is appropriate for E_c, so is

$$E_c^{+} = \frac{B_o(\rho^{+} - \rho_o^{+})^2}{2\rho_o^{+}\rho^{+2}} = \frac{B_o}{2\rho_o^{+}}\left[1 - \frac{1}{f(p_c^{+})}\right]^2$$

Accordingly we obtain

$$\frac{2E_c^+}{v^+} = \frac{B_o}{f(p_c^+)}\left[f(p_c^+) - 1\right]^2 \tag{3.37c}$$

Now Eqs. (3.37a), (3.37b) and (3.37c) suffice to determine Eq. (3.36b) or curve $v_o{}^+p_H{}^+v^+$ of Fig 3.3. The shock Hugoniot U = U(u) can be obtained by plotting data points (u, U) from $u = \sqrt{p_H^+(v_o^+ - v^+)}$ and $U = v_o^+\sqrt{p_H^+/(v_o^+ - v^+)}$.

3.9 Thomas-Fermi EOS for dense matter at ultra-high pressure

At pressure p > 10 Mb (10^7 atmospheres) all states of matter become a structureless assembly of nuclei and electrons and the electrons turn out to play a dominant role in compressive response. The Thomas-Fermi (TF) theory provides a high-pressure EOS on the basis of two assumptions: (a) the number density of electrons to be determined by the quantum statistics of a completely degenerate Fermi gas at T = 0°K, and (b) the electric charge to be approximated by Poisson's equation of classical electrostatics in spherical coordinates $\left(\nabla^2 = \frac{1}{r^2}\frac{d}{dr}\left(r^2\frac{d}{dr}\right)\right)$. It may be noted that the mathematics of TF theory is a bit obscure because several atomic constants have to be juggling around. Nevertheless we will here clarify the presentation tersely.

Let us first summarize the properties and EOS of a completely degenerate electron gas at T = 0°K:

Number density $n = \frac{N}{V} = \frac{8\pi}{3h^3}\Pi_o^3$ [cm^{-3}] (a1)

Fermi energy $\varepsilon_o = \frac{\Pi_o^2}{2m_-} = \frac{h^2}{2m_-}\left(\frac{3n}{8\pi}\right)^{\frac{2}{3}} = 5.842 \times 10^{-27} n^{2/3}$ [erg] (a2)

Degeneracy temperature $T_o = \varepsilon_o/k = 4.333 \times 10^{-11} n^{2/3}$ [°K]

(T > T_o non-degenerate, T < T_o degenerate)

$$\text{Equation of state } p_o = \frac{2E_o}{3V} = \frac{2}{5}n\varepsilon_o = \frac{h^2}{5m_-}\left(\frac{3}{8\pi}\right)^{\frac{2}{3}} n^{5/3} \quad \left[\frac{\text{dyn}}{\text{cm}^3}\right] \tag{3.38a}$$

Note that Eq. (3.38a) implies $E_O = 3N\varepsilon_O/5$ from Fermi-Dirac statistics. For symbols and constants see the list at the end of this section.

From the conservation of energy $\frac{\Pi_o^2}{2m_-} - e_-\phi = -e_-\phi_o$, we may re-write Eq. (a1) as

$$n = \left(8\pi/3h^3\right)\left[2m_-e_-(\phi - \phi_o)\right]^{3/2} \tag{b1}$$

Substituting this into the Poisson equation, we obtain

$$\nabla^2\phi = -4\pi n(-e_-) = \left(32\pi^2 e_-/3h^3\right)\left[2m_-e_-(\phi - \phi_o)\right]^{3/2}$$

$$\frac{1}{r^2}\frac{d}{dr}\left[r^2\frac{d}{dr}\left(\frac{Ze_-}{r}\psi + \varphi_o\right)\right] = \frac{32e_-\pi^2}{3h^3}\left(2m_-Ze_-^2\frac{\psi}{r}\right)^{3/2} \tag{b2}$$

$$\frac{Z}{\mu^3 x}\frac{d^2\psi}{dx^2} = \frac{32\pi^2}{3h^3}\left(\frac{2m_-Ze_-^2}{\mu}\frac{\psi}{x}\right)^{3/2} \quad \text{with } r = \mu x$$

$$\frac{d^2\psi}{dx^2} = \frac{\psi^{3/2}}{x^{1/2}} \quad \text{with } \frac{\mu^3}{Z}\left(\frac{32\pi^2}{3h^3}\right)\left(\frac{2m_-Ze_-^2}{\mu}\right)^{3/2} = 1 \tag{b3}$$

Thus Eq. (b3) can serve to determine the potential function

$$\phi - \phi_o = \frac{Ze_-}{r}\psi(r) = \frac{Ze_-}{\mu}\frac{\psi(x)}{x}$$

with boundary values $\phi - \phi_o = \frac{Ze_-}{r_o}$ at $r = 0$ and $f = -\frac{d\phi}{dr} = 0$ at $r = r_o$ (Note $V_o = 4\pi r_o^3/3$ = atomic volume and also $\mu = \left(h^2/m_-e_-^2\right)\left(9\pi^2/128Z\right)^{1/3}$ for reducing Eq. (b2) into Eq. (b3) in dimensionless form). In other words, we can seek the solution of Eq. (b3) with boundary values $\psi(0) = 1$ at $x = 0$ and $\psi(x_0) = x_0\psi'(x_0)$ at $x = x_0$ ($\psi' = d\psi/dx$):

$$\psi(x) = \sum_k b_k x^{k/2} = 1 + b_2 x + b_3 x^{3/2} + \ldots \tag{b4}$$

with $b_0 = \psi(0) = 1$, $b_1 = 0$, $b_2 = \psi'(0)$, $b_3 = 4/3$, $b_4 = 0$, $b_5 = 2b_2/5$, $b_6 = 1/3$, $b_7 = 3b_2/70$, etc. from direct substitution of $\psi(x)$ into Eq. (b3). Numerical results of $[x_0, \psi(x_0)]$ are tabulated elsewhere for arbitrary $V_o = 4\pi\mu^3 x_0^3/3$ which is pressure dependent. In this connection, we may re-write Eq. (b1) as

$$n_o = \frac{8\pi}{3h^3}\left(\frac{2mZe_-^2}{\mu x_o}\psi_o\right)^{3/2} = \frac{32Z^2}{9\pi^3 a_o^3}\left(\frac{\psi_o}{x_o}\right)^{3/2} \tag{b1'}$$

where $a_o = \hbar^2/m_- e_-^2$ stands for the Bohr radius of hydrogen atom. Substituting this into Eq. (3.38a), we get

$$p_o = \frac{\hbar^2}{5m_-}\left(\frac{3}{8\pi}\right) n_o^{5/3} = \frac{128}{45\pi^3}\left(\frac{4}{3\pi}\right)^{2/3}\frac{e_-^2}{a_o^4} Z^{\frac{10}{3}}\left(\frac{\psi_o}{x_o}\right)^{5/2}$$

$$= 1.524\text{x}10^{13} Z^{\frac{10}{3}}\left(\frac{\psi_o}{x_o}\right)^{5/2}\left[\frac{\text{dyne}}{\text{cm}^2}\right] \tag{3.38b}$$

which is the Thomas-Fermi EOS. Elsewhere Eq. (3.38b) is also written as

$$p_o = \frac{Z^2 e_-^2}{10\pi\mu^4}\left(\frac{\psi_o}{x_o}\right)^{5/2}$$

or

$$p_o V_o = \frac{2Z^2 e_-^2}{15\mu} x_o^{\frac{1}{2}} \psi_o^{\frac{5}{2}} \tag{b5}$$

On the other hand, let us introduce an approximate expression of Eq. (b4):

$$\psi_o = \frac{2.060 y_o}{(y_o - 0.2651)(1 + y_o)^{3.886}} \tag{b4'}$$

with $y_o = (x_o/12^{2/3})^{0.772}$ for $x_0 < 2.5$. Eq. (b4') is due to Sommerfeld whose original equation yields excellent results for $x_0 > 5$. For a given material with $\rho = m/N_oV_o$. (Avogadro's number $N_o = 6.0226 \times 10^{23}$), we have

$$\frac{ZV_o}{a_o^3} = \frac{Zm}{N_o a_o^3 \rho} = 11.2\frac{Zm}{\rho}$$

$$x_o = \left(\frac{3V_o}{4\pi\mu^3}\right)^{1/3} = 0.7\left(\frac{ZV_o}{a_o^3}\right)^{1/3} = 1.567\left(\frac{Zm}{\rho}\right)^{1/3}$$

$$y_o = \left(\frac{x_o}{12^{2/3}}\right)^{0.772} = 0.212\left(\frac{ZV_o}{a_o^3}\right)^{1/3} = 0.394\left(\frac{Zm}{\rho}\right)^{1/3}$$

Pertinent substitution of Eq. (b4') into Eq. (3.38b) gives

$$p_o = 81.45Z^{10/3}/\zeta^{0.19}\left(1.486\zeta^{0.257} - 1\right)^{2.50}\left(0.394\zeta^{0.257} + 1\right)^{9.72}, \text{ [Mb]} \tag{3.38c}$$

This form of Thomas-Fermi EOS has the practical merit with parameter $\zeta = Zm/\rho$ (Z = atomic number, m = atomic weight, and ρ = pressure-dependent density).

The Thomas-Fermi-Dirac (TFD) theory seeks to include the quantum effect of electron spins for further improvements with

$$n(x) = \frac{32Z^2}{9\pi^3 a_o^3}\left(\beta_o + \sqrt{\frac{\psi}{x}}\right)^3 \tag{b1''}$$

$$\frac{d^2\psi}{dx^2} = x\left(\beta_o + \sqrt{\frac{\psi}{x}}\right)^3 \tag{b3'}$$

$$p_oV_o = \frac{2Z^2e^2}{15\mu}x_o\left(\beta_o + \sqrt{\frac{\psi}{x}}\right)^4\left(\sqrt{\frac{\psi}{x}} - \frac{\beta_o}{4}\right) \tag{b5'}$$

with $\beta_o = (3/32\pi^2Z^2)^{1/3}$. Apparently Eqs. (b1"), (b3'), and (b5') reduce to Eqs. (b1'), (b3), and (b5) respectively by dropping β_0 (TF theory including no β_0).

EOS data from underground nuclear explosions indicate good agreement with TF results in the range 10Mb < p < 1,000Mb. Both TF and TFD theories are useful for preliminary inquiries of star's interior and ICF (inertial confinement fusion of deuterium and tritium). Yet more elaboration is needed for studies with high T.

Appendix - The following short list of symbols and atomic constants can serve to gain a better understanding of the TF and TFD theories.

$a_o = \hbar^2/m_- e_-^2 = 0.529 \times 10^{-8}$[cm] (Bohr's radius of H atom)

$b_k =$ constant coefficients of series solution $\psi = \sum_k b_k x^{k/2}$, $k = 1,2,3...$

$\beta_o = (3/32\pi^2 Z^2)^{1/3}$, a dimensionless parameter of TFD theory

$e_- = 4.803 \times 10^{-10}$[e.s.u.] (electron charge)

$h = 6.62 \times 10^{-27}$[erg - sec] (Planck quantum constant $\hbar$)

$k = 1.380 \times 10^{-16}$[erg/°K] (Boltzmann constant $\varepsilon = kT$)

$m =$ atomic weight of matter

$m_- = 9.108 \times 10^{-28}$[g] (electron mass)

$n = 8\pi\Pi_o^3/3h^3$ (number density of Fermi gas, $\Pi_o =$ Fermi momentum)

$N_o = 6.0226 \times 10^{23}$ (Avogadro's constant)

$p_o = 2n\varepsilon_o/5$ (EOS of completely degenreate electron gas,

$p_o \approx p_c =$ cold compression pressure of metal at T = 0°K)

$\mu = a_o\left(\frac{9\pi^2}{128Z}\right)^{1/3} = 0.8853 a_o Z^{-1/3} = 0.468 \times 10^{-8} Z^{-1/3}$[cm], a length scale to make $x = r/\mu$ dimensionless in $\psi(x)$.

$V_o = \frac{4}{3}\pi r_o^3 = \frac{4}{3}\pi\mu^3 x_o^3$ (atomic volume in TF and TFD equations of state).

$x_o = \left(\frac{3V_o}{4\pi\mu^3}\right)^{1/3} = 0.7\left(\frac{ZV_o}{a_o^3}\right)^{1/3} = 1.567\left(\frac{Zm}{\rho}\right)^{1/3}$, dimensionless atomic radius

Z = atomic number, ν = neutron number, $Z + \nu =$ mass number

$\zeta = \frac{Zm}{\rho}$[cm^3], the abbreviation for Eq. (3.38c)

Note that 1 atom contains Z electrons, Z protons, and ν neutrons (there can be more or less neutrons than protons, e.g. isotopes ${}^{1}_{1}H_{0}$, ${}^{2}_{1}H_{1}$, ${}^{3}_{1}H_{2}$ with $Z = 1$, $\nu = 0, 1, 2$). TF theory favors large atomic number Z for electric charges to be more or less continuously distributed as electron clouds surrounding the nuclei. Yet TF theory is still good for atoms with small atomic number Z at ultra-high pressure as more electrons are pooled together by compression.

3.10 Summary

For years EOS have been deeply rooted in the fertile soil of thermodynamics and statistical mechanics. In Section 3.2 we offer a theory for the construction of EOS with parameters γ, Γ, g, and j. Chapter 2 is already a comprehensible application of the γ-law gas for the development of basic SW theories. Now quantum ideal gases turn out to be a more general class of γ-law gases with parameters $\Gamma = g = \gamma\text{-}1$. In Section 3.4 the properties and EOS of phonons, photons, fermions, rotons, and bosons are described in terms of parameters (g, j) from the vantage point of thermodynamic simplicity. Degenerate gases can serve as a useful probe for the inquiry of extreme conditions and environments (high p, low T, or high T). The EOS of classical ideal gas and van der Waals fluids should be looked upon as complementary examples for the study of SW. The empirical EOS of Tait, Bridgman, and Murnaghan are useful for assessing the second-order compressibility of liquids and solids. Section 3.7 is especially worth pondering of the three versions of Grüneisen EOS plus several semi-analytical lattice potentials for solids. Eqs. (3.28b) and (3.33b) suffice to explore the volume dependence of $\Gamma = \Gamma(v)$ comprehensibly. In Section 3.8 we offer a short survey of the EOS for polymers, granular and porous material based on Grüneisen Γ and distention $\mu = \rho/\rho^{+}$. In the literature there appear a number of unconvinced papers which propose new EOS of porous materials, but our treatment of the topic is meant to be appropriate. Finally we seek to simplify the formulation of Thomas-Fermi and Thomas-Fermi-Dirac EOS for dense matter at ultra-high pressures. The reader is urged to revisit Section 3.9 again and over again until he grasps the essence of TF theory. The study of EOS will not end herewith, and the application of EOS is far-reaching. It should be borne in mind that all EOS, empirical and theoretical, have limited ranges of validity. Any viable EOS should prove its applicability for the right problem and environment. A few dozen of selected EOS are such examples for our purpose of SW study in this book.

Chapter 4 Some interdisciplinary SW and EOS

4.1 Hydraulic jumps and underwater shocks

Canal surges, flood waves, estuary bores, ocean breakers, surfs, and hydraulic jumps are all governed by these shock jump conditions:

$$y_1u_1 = y_2u_2$$
$$y_1u_1^2 + \frac{1}{2}gy_1^2 = y_2u_2^2 + \frac{1}{2}gy_2^2$$

Here y_i is the water depth, u_i the flow velocity with respect to the stationary shock front (i = 1 and 2 for the upstream and downstream respectively), and g the acceleration of gravity. Density is assumed constant for incompressible flow. Like the sound velocity and Mach number in compressible flow, the celerity of surface wave and Froude number are given by $c_i^2 = gy_i$ and $F_i = u_i/c_i$ $(i = 1, 2)$ respectively. Combining the above two equations, we obtain

$$\frac{1}{2}g(y_1^2 - y_2^2) = y_1u_1^2(y_1 - y_2)/y_2 = y_2u_2^2(y_1 - y_2)/y_1$$

which gives

$$u_1^2 = \frac{1}{2}gy_2\left(1+\frac{y_2}{y_1}\right),\quad F_1^2 = \frac{1}{2}\frac{y_2}{y_1}\left(\frac{y_2}{y_1}+1\right),\quad \frac{y_2}{y_1} = \frac{1}{2}\left(\sqrt{1+8F_1^2} - 1\right)$$
$$u_2^2 = \frac{1}{2}gy_1\left(1+\frac{y_1}{y_2}\right),\quad F_2^2 = \frac{1}{2}\frac{y_1}{y_2}\left(\frac{y_1}{y_2}+1\right),\quad \frac{y_1}{y_2} = \frac{1}{2}\left(\sqrt{1+8F_2^2} - 1\right)$$

Moreover we can count the energy difference

$$\Delta\mathcal{E} = \left(\frac{1}{2}u_2^2 + gy_2\right) - \left(\frac{1}{2}u_1^2 + gy_1\right) = \frac{g}{4y_1y_2}(y_2 - y_1)^3$$

From these we may specify the properties of a hydraulic jump as follows:

$y_2 > y_1$, $F_1 > 1$, $u_1 > c_1$ (supercritical); $y_1 < y_2$
$F_2 < 1$, $u_2 < c_2$ (subcritical); and $\Delta\mathcal{E} > 0$.

All these correspond to the normal shock exactly. See Section 1.7 and Eq. (1.14a)

It should be noted that a hydraulic jump is an isothermal, normal shock in the incompressible flow of water under the influence of gravity. Accordingly it is also known as gravity shock (vs. gasdynamic shock due to nonlinearity and compressibility). The hydrostatic $p = 1/2\, \rho g y^2$ may be looked upon as the isothermal EOS for the gravity shock, and $\Delta\mathcal{E}$ goes to the irreversible dissipation of turbulence therewith. When the shock front moves against the stream, we have a negative bore or surge. But it is not a negative shock or rarefaction shock! Yet another phenomenon is the dam break or flush flood which involves hydraulic jump and release waves just like the waves system of gasdynamic shock tube (see Section 5.1). Note also that the analysis of hydraulic jump needs to be modified if the flow involves elevation and/or volume gradients.

Underwater shock waves are more or less similar to blast waves in the air, which are already discussed in Section 2.3. Here the phenomenology of underwater explosion is worth noting. When a spherical charge is initiated in deep water, the detonation wave spreads to consume all the explosive and generates a hot gas bubble, the high pressure of which instantly sends out a spherical SW. The propagation of this primary SW may encounter obstacles or boundaries, resulting in reflection, refraction, and diffraction to cause damage, cratering, or cavitation. On the other hand, the gas bubble expands with rarefaction waves and contracts by the exterior hydrostatic pressure in a few cycles. Each time when the gas bubble contracts to the limit of implosion, a pressure pulse sets out from the center of implosion (note that this is much weaker than the primary SW). Eventually the pulsating gas bubble will emerge from the free surface as dome and plume spray. Other outcomes such as surface waves or breakers are widespread.

It is of interest to inquire into two major factors of the above account, namely, the primary SW and the pulsation of the gas sphere. Since the detonation products and water may be considered as γ-law fluids with $\gamma = 3$ and 7 respectively, the model of blast waves and scaling laws are applicable to deep underwater SW (see Eqs. (2.25)

- (2.29) and the cube-root scaling) by adopting the pertinent constants γ, Y, ρ_0, and B. Underwater SW propagate with diminishing strengths due to the divergence of the shock front. Thus we have the decay rate

$$\frac{dp_H}{dt} = -\frac{6p_H}{5t} \tag{4.1a}$$

in comparison with the empirical results

$$\frac{dp_H}{dt} = -\frac{p_H}{\theta} \tag{4.2a}$$

$$p_H = \hat{p}e^{-t/\theta} \tag{4.2b}$$

$$\hat{p} = c\left(W^{1/3}/R\right)^n \tag{4.2c}$$

Here we have $\hat{p}$ for the peak shock pressure at distance R from the center of explosive (its linear dimension being proportional to the cube root of its weight W), with θ, c, and n denoting empirical constants. Note that Eq. (4.2a) implies additional effects of viscosity and density. Assuming the strong shock limit $p_H \propto U^2$ in connection with Eq. (4.2a), we may write

$$\frac{dU}{dt} = -\frac{U}{2\theta} \tag{4.2d}$$

or

$$U = \hat{U}e^{-t/2\theta} \tag{4.2e}$$

to compare with the result of Eqs. (2.25) and (2.27), namely

$$U = \left(\frac{8B^2}{125}\right)^{1/5} t^{-3/5} \tag{4.1b}$$

Of course all these are only a simplified representation of the underwater SW, while numerical simulations may provide better results at a price.

The equation of motion for the water around the gas bubble may be approximated by

$$\rho\left(r\ddot{r} + 3\dot{r}^2/2\right) = p \tag{4.3a}$$

which is a result of combining the continuity and momentum equations of incompressible hydrodynamics. Note $\dot{r} = dr/dt$ and $\ddot{r} = d^2r/dt^2$. For the initial state of the gas bubble (p_o, $V_o = 4\pi r_o^3/3$), we have $p/p_o = \left(V_o/V\right)^\gamma = \left(r_o/r\right)^{3\gamma}$. Multiplying both sides of Eq. (4.3a) by $r^2\dot{r}$ and letting $p_o/\rho = c_o^2$, we obtain

$$\frac{d}{dt}\left(r^3\dot{r}^2\right) = \frac{2c_o^2 r_o^{3\gamma}}{r^{3\gamma-2}}\frac{dr}{dt}$$

and hence

$$\dot{r}^2 = \frac{2c_o^2}{3(\gamma-1)}\left[\left(\frac{r_o}{r}\right)^3 - \left(\frac{r_o}{r}\right)^{3\gamma}\right] \tag{4.3b}$$

Let $r = (1+x)r_o$ and $dr = r_o dx$ for $\gamma = 4/3$. Then Eq. (4.3b) reduces to

$$r_o dx = c_o\sqrt{2x}(1+x)^{-2}dt$$

which is integrable to give

$$c_o t/r_o = \left(1 + 2x/3 + x^2/5\right)\sqrt{2x} \tag{4.3c}$$

For example let $x = 1$ for $r_o = 50$ cm, $p_o = 1{,}000$ atm., $c_o = \sqrt{p_o/\rho} = 3.16 \times 10^4$ cm/sec. Eq. (4.3c) gives $t = 0.004$sec for the gas bubble expansion from r_0 to $r = 2r_0$. The radial velocity $r = c_o/\sqrt{8} = 1.117 \times 10^4$ cm/sec according to Eq. (4.3b). Note that our computation is only good for the early stage of bubble expansion. Eq. (4.3b) must be modified to account for the oscillatory expansion and contraction of the gas bubble.

Considerable effort had been focused on the testing and research of underwater explosion for weapons technology during the 1940's. Yet another development is the meaningful application of underwater SW to shatter kidney stones during the late 1970's. In Section 5.6 we shall examine this aspect further.

4.2 Negative shocks and SW involving phase transition

For a given EOS the sign of $(\partial p/\partial v)_s$ and $(\partial^2 p/\partial v^2)_s$ will determine the formation, propagation, and stability of SW uniquely. Apparently the sound velocity requires $(\partial p/\partial v)_s < 0$, and it is necessary to have $(\partial^2 p/\partial v^2)_s > 0$ for compression shocks with $ds > 0$. Also rarefaction waves will have to propagate continuously with $ds = 0$. So far the majority of shock phenomena and EOS falls in this category of positive SW. On the other hand, negative shocks are equally legitimate for EOS with $(\partial p/\partial v)_s < 0$ and $(\partial^2 p/\partial v^2)_s < 0$. See Eq. (1.14a). Negative shocks appear as a result of rarefaction waves coalescing into a sudden expansion with $ds > 0$, whereas supersonic compression waves cannot form a steep wave front (now propagating continuously with $ds = 0$). Such behavior is peculiar only with fluids near the critical point. Note that both positive and negative SW satisfy the same jump conditions as described by Eqs. (1.1) - (1.7) and (1.14a), and their fundamental properties are summarized in Section 1.7.

Let us now explore negative SW further using the van der Waals EOS which is adaptable to a spectrum of behavior and phases with $(\partial p/\partial v)_s \lessgtr 0$, $(\partial^2 p/\partial v^2)_s \lessgtr 0$, and a critical point defined by $(\partial p/\partial v)_T = (\partial^2 p/\partial v^2)_T = 0$. Note that most EOS have $(\partial p/\partial v)_s < 0$ and $(\partial^2 p/\partial v^2)_s > 0$ otherwise monotonously (viz. impossible for negative shocks). Introducing $\Pi = p/p_c$, $\phi = v/v_c$, and $\Theta = T/T_c$, we may re-write Eqs. (3.19), (3.22), and (1.14a) in dimensionless forms as

$$\left(\Pi + \frac{3}{\phi^2}\right)(3\phi - 1) = 8\Theta \tag{4.4}$$

$$\Pi'' = \left(\partial^2 \Pi/\partial\phi^2\right)_\theta = 144\Theta(3\phi - 1)^{-3} - 18\phi^{-4} \tag{4.4a}$$

$$\frac{\Pi_2}{\Pi_1} = \frac{\left(\phi_1 - \frac{1}{3}\right) - \frac{1}{2}(\lambda - 1)(\phi_2 - \phi_1) + \frac{1}{\Pi_1}\left(\frac{1}{\phi_2} - \frac{1}{\phi_1}\right)\left(\frac{1}{\phi_2} + \frac{1}{\phi_1} + 3\lambda - 6\right)}{\left(\phi_2 - \frac{1}{3}\right) + \frac{1}{2}(\lambda - 1)(\phi_2 - \phi_1)} \tag{4.5}$$

$$(s_2 - s_1)/R = (\partial^2\Pi/\partial\phi^2)_s(\phi_1 - \phi_2)^3/32\Theta_1 \tag{4.6}$$

respectively. From Eqs. (4.4) and (4.4a) we obtain these data points for $\Theta = 1$:

ϕ	1.00	1.10	1.85	1.90
Π_c	1.00	0.999	0.882	0.871
Π_c''	0.00	−0.468	−0.008	+0.006

which mark a sector of the critical isotherm with $\partial^2\Pi/\partial\phi^2 < 0$. Now that retrograde fluids have $c_v >> R$ near the critical point (e.g. $c_v/R = 36.82$ for n-octane C_8H_8 and $c_v/R = 109$ for fluorinated ether E-4 $C_{14}F_{29}HO_4$), their isentropes with $\lambda = 1 + R/c_v \approx 1$ may be treated as isotherms. Accordingly we may use $(\partial^2p/\partial v^2)_S \approx (\partial^2p/\partial v^2)_T < 0$ for the computation of negative SW.

Let us consider an example by using Eqs. (4.4) - (4.6). Given $c_v/R = 50$ ($\lambda = 1.02$), $\phi_1 = 1.05$ and $\Theta_1 = 1.02$, we obtain $\Pi_1 = 1.074$ and $\Pi'' = -0.030$. The isotherm for $\Theta_1 = 1.02$ has $\Pi_2'' = +0.034$ for $\phi_2 = 1.90$. So we try $\phi_2 = 1.80$ with $\Pi_2'' = +0.009$ which should be again excluded for a negative shock. For $\phi_2 = 1.75$ we obtain $\Pi_2'' = -0.006$, $\Pi_2 = 0.830$, $\Theta_2 = 0.961$, and $(s_2 - s_1)/R = 0.00032$ from Eqs. (4.4a), (4.5), (4.4), and (4.6) respectively. Thus we have $v_2/v_1 = 1.67$, $p_2/p_1 = 0.773$, $T_2/T_1 = 0.942$, and $ds > 0$ for a negative shock as a sudden expansion ($v_2 > v_1$) with pressure drop ($p_2 < p_1$), temperature fall ($T_2 < T_1$), and entropy production ($s_2 > s_1$). Note here $c_v >> R$ and hence $(\partial^2p/\partial v^2)_S \approx (\partial^2p/\partial v^2)_T < 0$. For the limit of shock expansion, we need two equations. For simplicity we let $\lambda = \phi_1 = \Pi_1 = \Theta_1 = 1$ and re-write Eq. (4.5) as

$$\Pi_h = (8\phi^2 - 9\phi + 3)/\phi^2(3\phi - 1)$$

to approximate the shock EOS. From Eqs. (4.4a) and (4.4) we may write

$$\Pi_f = \left(6\phi^2 - 6\phi + 1\right)/\phi^4$$

for the locus of inflection points $(\partial^2\Pi/\partial\phi^2)_\theta = 0$. Let $\Pi_h = \Pi_f$ or

$$8\phi^4 - 27\phi^3 + 27\phi^2 - 9\phi + 1 = 0$$

which has two roots $\phi = 1$ ($\Pi_h = \Pi_f = 1$) and $\phi = 1.878$ ($\Pi_f = \Pi_h = 0.79$). The second root determines the limit of the negative shock. Compare the difference between this and the strong limit of positive shock as given by Eq. (3.22a).

As mentioned earlier, the van der Waals EOS is adaptable to a variety of regions, states (stable, metastable, unstable), and phases (liquid, vapor, mixture). The initial state (p_1, v_1) and the sign of $(\partial^2 p/\partial v^2)_T$ turn out to unfold a new catalog of shock waves in the van der Waals fluid. Thus all positive shocks are featured with $(\partial^2 p/\partial v^2)_T > 0$, such as shock compression of liquid or vapor in 1-phase region. Liquefaction shocks are positive, involving the transformation of vapor into mixture or saturated liquid in 2-phase region (note that condensation shock is the term for water droplets formed in moisture air or wet steam by shock compression in nozzle flow). In regions of $(\partial^2 p/\partial v^2)_T < 0$ three types of negative shocks have been detected: (a) rarefaction shocks with supersaturated-to-dry vapor transformations in 1-phase region, (b) rarefaction shocks in 2-phase region convert a near critical vapor into a mixture, and (c) evaporation shocks transform a 2-phase mixture into a dry vapor. Along with the Hugoniot curve, the piecewise continuous isentrope exhibits kinks [i.e. $(\partial p/\partial v)_S$ discontinuities] across the binodal (or saturation line), and so phase-transition shocks split with two sound speeds. Yet another 2-phase shock phenomenon is the vapor explosion (e.g. explosive boiling of cold liquid in superheated liquid) which is the subsonic branch of vaporization Hugoniot in analogy to the deflagration of explosive. Note that dynamic phase transitions are realized in microseconds by positive or negative shocks (cf. the sluggish, static phase transformations by isotherms). It is the latent heat (of large c_v) and many molecular degrees of freedom that promote catalytic action or relaxation in dynamic phase transitions of retrograde fluids. All these shock phenomena have been observed in shock-tube experiments, but their computation is rather laborious.

4.3 Detonation waves

At certain pressure and temperature combustible gases may react explosively. Likewise dust explosions occur illimitably. Condensed liquid and solid explosives are more powerful for industrial and military applications. Here we seek to investigate the theoretical aspects of these phenomena. Detonation is the rapid and violent decomposition of high explosives (HE) initiated by SW; the shock and chemical reaction interact and support each other so that a steady propagation of both is maintained at a constant velocity D; and hence a detonation wave is a reactive SW with the von Neumann spike (viz. the shock front) adjoined by the reaction zone (viz. the shock thickness to mark out the Chapman-Jouguet (C-J) plane). Thus other SW with no chemical reactions involved may be regarded as inert SW, and both reactive and inert SW differ considerably in their formation and propagation.

Let us depict in Fig. 4.1 the Hugoniot curves of unreacted HE and completely reacted detonation products (DP) by *ohnn'* and *wjs* respectively. When a HE is

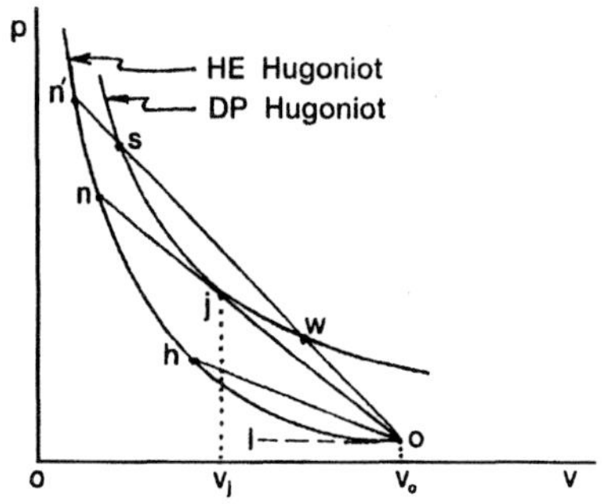

Fig. 4.1 Hugoniot curves for C-J detonation.

compressed from initial state *o* to state *n* (the von Neumann spike), the pressure and temperature are high enough to initiate reaction which completes instantly along *nj*. Note that the secant *on* of HE Hugoniot is the Rayleigh line just tangent to DP Hugoniot at point *j* (the C-J state of detonation). The Rayleigh line defines the shock impedance $(\rho_0 D)^2 = (p_j - p_0)/(v_0 - v_j) = \tan\theta$, and the tangent *oj* defines the acoustic impedance $-(\partial p/\partial v)_s = (\rho_j c_j)^2 = \gamma p_j/v_j$. These two impedances turn out to be equal because of the geometrical tangency. This is the gist of C-J steady detonation. In Fig. 4.2 we sketch the steady detonation with von Neumann spike *on*, Chapman-Jouguet plane *cj*, and reaction zone *oc*. Since the shock thickness is of magnitude at

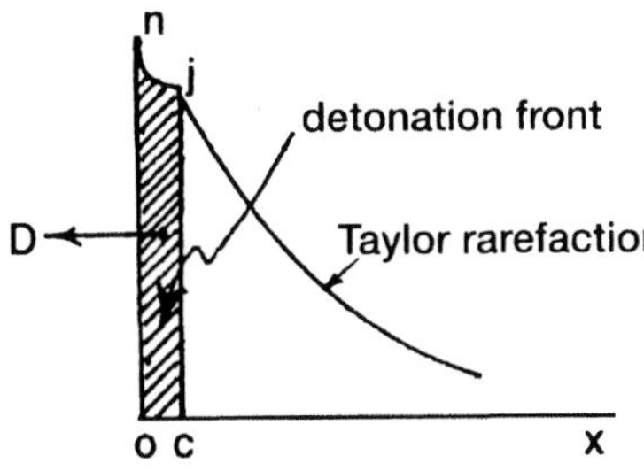

Fig. 4.2 Steady detonation wave.

micro-level, chemical reaction may be considered complete instantly (along segment *nj* in Figs. 4.1 and 4.2) to yield heat of reaction Q in support of the C-J detonation. It is of relevance to clarify the representation of Fig. 4.1 further. Let $\theta = \angle loj$ and $\tan\theta = (\rho_0 D)^2$ as mentioned earlier. For a given shock whose Rayleigh line *oh* does not intersect DP Hugoniot, we have $(\rho U_h)^2 = (p_h - p_0)/(v_0 - v_h) = \tan\angle loh$. Accordingly this shock will not initiate a detonation with $p_h < p_n$ (von Neumann spike) and $U_h < D$ (because of $\angle loh < \theta$). On the other hand, Rayleigh line on' with $\angle los > \theta$ represents a reactive SW which cuts DP Hugoniot between *s* and *w*. These two states refer to detonations of different behavior. State *w* is a weak detonation whose downstream supersonic flow ($u_2 > c_2$) violates the shock jump condition ($u_2 < c_{2'}$). Such a detonation with $p_w > p_j$ and $U_w > D$ is unstable or unsupported despite shock initiation at $p_{n'}$. State *s* is a strong or overdriven detonation with $p_s > p_j$ and $U_s = U_w > D$ (also downstream subsonic flow $u_2 < c_2$). Overdriven detonations are achieved in convergent detonation waves or by primary booster ($p_s > p_j$).

A simplified theory may be formulated for the C-J detonation based on the EOS of γ-law gases. Let DP state be denoted by (p_j, ρ_j, E_j, u_j, c_j). From the impedance and continuity relations, we may write

$$\rho_j c_j = \rho_o D = \rho_j (D - u_j)$$

and hence

$$D = u_j + c_j \tag{4.7}$$

In view of $p_0 << p_j$, we may re-write

$$p_j/(v_o - v_j) = (\rho_o D)^2 = (\rho_j c_j)^2 = \gamma p_j / v_j$$

which leads to

$$\rho_j/\rho_o = (\gamma+1)/\gamma \tag{4.8}$$

and

$$c_j = \gamma u_j \tag{4.9}$$

From the momentum relation and Eqs. (4.7) and (4.9), we deduce

$$p_j = \rho_o u_j D = \rho_o D^2/(\gamma+1) \tag{4.10}$$

Since E_0 is negligible to compare with Q, the energy equation may be written as

$$E_j = p_j v_j/(\gamma-1) = \frac{1}{2}u_j^2 + Q \tag{4.11}$$

By substitution of Eqs. (4.7) - (4.10), we have

$$Q + \frac{1}{2}\left(\frac{D}{\gamma+1}\right)^2 = \frac{\gamma D^2}{(\gamma-1)(\gamma+1)^2} = E_j \tag{4.11a}$$

and hence

$$D^2 = 2Q(\gamma^2 - 1) \tag{4.12}$$

$$E_j = 2\gamma Q/(\gamma+1) \tag{4.11b}$$

Yet another formulation is worth noting, which dictates Eq. (4.12) to give the minimum shock velocity for the steady propagation of C-J detonation. Let us derive the shock Hugoniot U=U(u) for the reactive SW by using all jump conditions together:

$$Q+\frac{u^2}{2}=\frac{pv}{\gamma-1}=\frac{\rho_o uU/\rho}{\gamma-1}=\frac{uU-u^2}{\gamma-1}$$

which gives

$$U=\left(\frac{\gamma+1}{2}\right)u+(\gamma-1)\frac{Q}{u} \tag{4.13}$$

Now we have

$$\frac{dU}{du}=\frac{\gamma+1}{2}-\frac{\gamma-1}{u^2}Q=0$$

$$u_{\min}=\sqrt{2Q\left(\frac{\gamma-1}{\gamma+1}\right)}$$

and

$$U_{\min}=\sqrt{2Q(\gamma^2-1)}$$

which is exactly Eq. (4.12). From this we may reverse the derivation of Eqs. (4.7) - (4.12) without reference to the tangent condition of C-J point.

The foregoing discussion is meant to be a concise introduction to the theory of reactive SW and C-J detonation. For analytical simplicity we have used only one EOS, namely, the γ-law gas for DP. The constant γ is to be given by

$$\gamma=\frac{\rho_o D^2}{p_j}-1 \tag{4.10a}$$

with experimental results. For solid HE γ has the order of magnitude 3. thus we mention

TNT (trinitrotoluene $C_7H_5N_3O_6$), ρ_o = 1.64 g/cm^3, D = 6.95 km/sec, p_j = 190 kb, and $\gamma = 3.16$.

RDX (cyclotrimethylene trinitramine $C_3H_6N_6O_6$), ρ_o = 1.80 g/cm^3, D = 8.75 km/sec, p_j = 347 kb, and $\gamma = 2.98$.

Comp B (64% RDX/36% TNT), ρ_o = 1.71 g/cm^3, D = 8.03 km/sec, p_j = 294 kb, and γ = 2.76.

Note that, in accurate analysis of reactive SW, we need two EOS: one for the unreacted HE and the other for DP. Apparently the former is simply a matter of choice from those of Chapter 3, but the latter is not so, because DP are a mixture of gases such as H_2O, CO_2, NO_2, OH, CO, etc. with variable composition. The EOS for such a gas mixture is too complex to formulate, and so we enlist a few of the more practical EOS below.

With $a = 0$ the van der Waals EOS becomes

$$p(v-b) = \mathrm{RT} \tag{3.19c}$$

which is also known as the Abel EOS for a modified γ-law gas. In view of the co-volume $b = 4nvv_m$, the Abel EOS turns out to uncover several useful EOS for DP. Here n is the number density of gas molecules, and the volume of a single molecule is $v_m = 4\pi r^3/3$ (r being the molecular radius). Let the packing fraction of molecules by $y = nv_m$. Then Eq. (3.19c) becomes

$$p = \frac{\mathrm{RT}}{v(1-4y)} \approx \left(\frac{\mathrm{RT}}{v}\right)\frac{1-y^3}{(1-y)^4} = \frac{\mathrm{RT}(1+y+y^2)}{v(1-y)^3} \tag{4.14}$$

which is the Percus-Yevick EOS for DP gases at high T. A similar EOS is given by Carnahan and Starling

$$\frac{pv}{\mathrm{RT}} = \frac{1+y+y^2-y^3}{(1-y)^3} \tag{4.15}$$

for monatomic gases though.

It is of analytical interest to note the virial EOS

$$\frac{pv}{\mathrm{RT}} = 1 + \frac{\mathrm{B}}{v} + \frac{\mathrm{C}}{v^2} + \ldots = 1 + x + \frac{\mathrm{C}}{\mathrm{B}^2}x^2 + \ldots = 1 + x + \beta x^2 + \ldots \approx 1 + xe^{\beta x}$$

with $x = \mathrm{B}/v$ and $\beta = \mathrm{C}/\mathrm{B}^2$ to be fitted empirically. Becker, Kistiakowsky, and Wilson (BKW) introduce four parameters (α, β, θ, κ) in $\mathrm{B} = \kappa/(\mathrm{T} + \theta)^\alpha$ and $x = \kappa/v(\mathrm{T} + \theta)^\alpha$ for

$$\frac{pv}{\mathrm{RT}} = 1 + xe^{\beta x} \tag{4.16}$$

This is the BKW equation of state for DP gases at high p. For TNT with $\rho_0 = 1.64$, it is appropriate to use $\beta = 0.096$, $\kappa = 12.69$, $\theta = 400°\mathrm{K}$, and $\alpha = 0.5$. Also it is suitable for RDX ($\rho_0 = 1.80$) to have $\beta = 0.16$, $\kappa = 10.90$, $\theta = 400°\mathrm{K}$, and $\alpha = 0.5$. At Los Alamos National Laboratory the BKW has been incorporated in a computer code named TIGER to expedite calculations. Yet another EOS has become popular in the community of detonation research. At Lawrence Livermore National Laboratory Jones, Wilson, and Lee (JWL) introduce

$$p_s = \mathrm{A}e^{-\mathrm{R}_1 v} + \mathrm{B}e^{-\mathrm{R}_2 v} + \mathrm{C}v^{-(\omega+1)}$$

with six parameters (A, B, C, R_1, R_2, ω) to be fitted empirically for the C-J insentrope. From this we obtain

$$E_s = -\int p_s dv = \frac{\mathrm{A}}{\mathrm{R}_1}e^{-\mathrm{R}_1 v} + \frac{\mathrm{B}}{\mathrm{R}_2}e^{-\mathrm{R}_2 v} + \frac{\mathrm{C}}{\omega}v^{-\omega}$$

$$p - p_s = \frac{\omega}{v}(\mathrm{E} - \mathrm{E}_s)$$

and hence

$$p = \mathrm{A}\left(1 - \frac{\omega}{\mathrm{R}_1 v}\right)e^{-\mathrm{R}_1 v} + \mathrm{B}\left(1 - \frac{\omega}{\mathrm{R}_2 v}\right)e^{-\mathrm{R}_2 v} + \frac{\omega}{v}\mathrm{E} \tag{4.17}$$

which is the JWL equation of state. For solid HE it is essentially the Grüneisen EOS with $\omega = \Gamma$. Of course it is also used for DP with different data fit. Thus far we have provided only a bird's-eye view of reactive SW and EOS without looking into the theory of deflagration.

4.4 Ionizing shocks

Shock compression of gases at high T may cause dissociation, ionization, and radiation. When ionized gases flow in magnetic fields, the Lorentz force comes into play which changes the pattern and behavior of SW remarkably. It is too complicated to take all these effects into account as a comprehensive study. Let us first consider the ionizing shock in hydrogen with Eqs. (3.11f) and (3.12), excluding dissociation, radiation, and magnetic fields. Eliminating δ from these two equations, we may write

$$(p - k\mathrm{T}\rho/m_\mathrm{H})^2 = k\mathrm{T}f(\mathrm{T})(2k\mathrm{T}\rho/m_\mathrm{H} - p) \tag{4.18}$$

as a compact EOS for the ionized hydrogen with

$$f(\mathrm{T}) = (m_e k\mathrm{T}/2\pi\hbar^2)^{3/2} \exp(-\mathrm{I}/k\mathrm{T}) \tag{4.18a}$$

Now the model of γ-law gas is slightly modified to have

$$\mathrm{E} = \frac{pv}{\gamma - 1} + \frac{\delta \mathrm{I}}{m_\mathrm{H}} \quad (0 \le \delta \le 1) \tag{4.19}$$

$$\mathrm{E}_1 = \frac{p_1 v_1}{\gamma - 1} \quad (\delta_1 = 0 \text{ before ionization}) \tag{4.19a}$$

$$\mathrm{E}_2 = \frac{p_2 v_2}{\gamma - 1} + \frac{\delta \mathrm{I}}{m_\mathrm{H}} \quad (\delta_2 = \delta \text{ depending upon } \mathrm{T}_2) \tag{4.19b}$$

Substituting Eqs. (4.19a) and (4.19b) into Eq. (1.7), we deduce

$$\frac{p_2}{p_1} = \frac{(\gamma+1)v_1 - (\gamma-1)v_2 - (2\delta I/kT_1)(\gamma-1)v_1}{(\gamma+1)v_2 - (\gamma-1)v_1} \tag{4.20}$$

as the shock EOS for the ionizing shock. From Eq. (3.11f) we have

$$\delta = \frac{m_H p_2 v_2}{kT_2} - 1 \tag{3.11f'}$$

Substituting this into Eq. (4.20) and rearranging, we obtain explicitly

$$\frac{p_2}{p_1} = \frac{[(\gamma+1)+(\gamma-1)(2I/kT_1)]v_1 - (\gamma-1)v_2}{[(\gamma+1)+(\gamma-1)(2I/kT_2)]v_2 - (\gamma-1)v_1} \tag{4.20a}$$

to complete our formulation.

The foregoing equations can serve to determine the shocked state (p_2, v_2, T_2, δ) yet a numerical solution for Eqs. (4.20a) and (4.18) is not quite straightforward because they contain three unknowns (p_2, v_2, T_2). So assuming a temperature T_2 for trial and putting p_2, $\rho_2 = 1/v_2$, and T_2 in Eqs. (4.18) and (4.18a), we can solve the simultaneous Eqs. (4.20a) and (4.18). Then δ is determined by Eq. (3.11f′). A simple computer program can deal with these manipulations speedily. Alternatively we may attempt first assuming an arbitrary degree of ionization (say, δ = 1% etc.) in Eq. (4.20) to solve p_2/p_1 in terms of v_1/v_2. The shock temperature is then obtained from

$$T_2 = m_H p_2 v_2 / k(1+\delta) \tag{3.11f''}$$

It is worth noting that the strong and weak shock limits are given by

$$\frac{\rho_2}{\rho_1} = \frac{\gamma+1}{\gamma-1} + \frac{2I}{kT_2} \quad (p_2 >> p_1) \tag{4.20b}$$

and

$$\frac{p_2}{p_1} = \frac{1+(\gamma-1)I/kT_1}{1+(\gamma-1)I/kT_2} \quad (v_2 \approx v_1 \text{ but } T_2 \gtrsim T_1) \tag{4.20c}$$

respectively. All results can be further simplified with $\gamma = 5/3$ for monatomic gases. Also for the special case $\delta = I = 0$, Eqs. (4.20) and (4.20a) reduce to Eq. (2.1a).

4.5 Radiating shocks

For a γ-law gas at very high T, there exists a radiation pressure component $p_r = bT^4$ which must be included in $P = p_g + p_r$ (b being a constant and $p_g = \rho RT$). Likewise the enthalpy equation is

$$h = h_g + h_r = \frac{\gamma}{\gamma - 1}\frac{p_g}{\rho} + \frac{4p_r}{\rho}$$

implying $\gamma_r = 4/3$ (see quantum ideal gas, photons, of Chapter 3). With these EOS, we may write the shock-jump equations below.

$$\rho_1 u_1 = \rho_2 u_2 \tag{1.1}$$

$$p_1 + \rho_1 u_1^2 = (p_2 + p_r) + \rho_2 u_2^2 \tag{1.2}$$

$$\left(\frac{\gamma}{\gamma - 1}\right)\frac{p_1}{\rho_1} + \frac{u_1^2}{2} = \left[\frac{\gamma}{\gamma - 1}\frac{p_2}{\rho_2} + \frac{4p_r}{\rho_2}\right] + \frac{u_2^2}{2} \tag{1.3}$$

where it is understood with no radiation components in state 1.

In order to solve this problem analytically, we seek to introduce these symbols: $x = \rho_2/\rho_1$, $y = p_2/p_1$, $\mu = u_2/u_1$, $\zeta = p_r/p_1 = bT_2^4/p_1 = (b/p_1)(p_2/\rho_2 R)^4 = (\alpha y/x)^4$ with $\alpha^4 = bp_1^3/\rho_1^4 R^4$, and $\phi^2 = \rho_1 u_1^2/p_1$. Now Eqs. (1.1) - (1.3) may be simplified approximately ($\approx$) as

$$\mu x = 1 \tag{1.1a}$$

$$y + \zeta = \phi^2(1 - \mu) + 1 \approx \phi^2(1 - \mu) \tag{1.2a}$$

$$\left(\frac{\gamma}{\gamma-1}\right)\frac{y}{x}+\frac{4\zeta}{x}=\frac{\gamma}{\gamma-1}+\frac{1}{2}\phi^2\left(1-\mu^2\right)\approx\frac{1}{2}\phi^2(1-\mu)(1+\mu) \tag{1.3a}$$

Re-writing Eq. (1.2a) with Eq. (1.1a) and (1.3a) we have

$$(y+\zeta)(1+\mu)=\left(\frac{2\gamma}{\gamma-1}y+8\zeta\right)\mu \tag{1.2b}$$

To simplify Eq. (1.2b) further, we may write

$$(y+\zeta)+(y+\zeta)\mu=\left[\left(1+\frac{\gamma+1}{\gamma-1}\right)y+8\zeta\right]\mu=\left[\beta y+7\zeta+(y+\zeta)\right]\mu$$

$$(\beta y+7\zeta)\mu=y+\zeta$$

and hence

$$y=\zeta\left(\frac{7\mu-1}{1-\beta\mu}\right)=\zeta\left(\frac{7-x}{x-\beta}\right) \tag{1.2c}$$

with $\beta=(\gamma+1)/(\gamma-1)$ as before. Substituting $\zeta=(\alpha y/x)^4$ in Eq. (1.2c) gives

$$y=\left(\frac{x}{\alpha}\right)^{4/3}\left(\frac{x-\beta}{7-x}\right)^{1/3}=\frac{(R\rho_1 x)^{4/3}}{p_1 b^{1/3}}\left[\frac{(\gamma-1)x-(\gamma+1)}{7(\gamma-1)-(\gamma-1)x}\right]^{1/3} \tag{1.2d}$$

or

$$p_2=(R\rho_2)^{4/3}\left[\frac{(\gamma-1)\rho_2-(\gamma+1)\rho_1}{b(\gamma-1)(7\rho_1-\rho_2)}\right]^{1/3} \tag{4.21}$$

This is the Hugoniot EOS for the γ - law radiating gas. Note $\phi^2=\rho_1 u_1^2/p_1=\gamma M_1^2$ which gives the shock Mach number

$$M_1=\sqrt{\frac{y+\zeta}{\gamma(1-\mu)}}=\sqrt{\frac{xy}{\gamma(x-1)}\left(\frac{7-\beta}{7-x}\right)} \tag{4.22}$$

by using Eqs. (1.1a), (1.2a), and (1.2c). Substituting all symbols (x, y, ζ, μ, β) in Eq. (4.22), we may write explicitly

$$M_1^2 = \frac{2}{\gamma}\left(\frac{3\gamma-4}{\gamma-1}\right)\frac{p_2}{p_1}\left(\frac{\rho_2}{\rho_2-\rho_1}\right)\left(\frac{\rho_1}{7\rho_1-\rho_2}\right) \tag{4.22a}$$

It is now of interest to consider Eq. (1.2c) with $x = \beta$ and $\zeta < y \to \infty$. That is to say the strong shock limit with $\rho_2/\rho_1 = (\gamma+1)/(\gamma-1) = 4$ for $\gamma = 5/3$, $p_r < p_2 \to \infty$ for the ideal gas to have insignificant radiation (then $\phi \to \infty$ and $M_1 \to \infty$, too). On the other hand, Eq. (1.2c) gives $x = 7$ and $y < \zeta \to \infty$. This is the strong shock limit dominated by radiation ($p_2 < p_r \to \infty$, $\rho_2/\rho_1 = (\gamma+1)/(\gamma-1) = 7$ for photon gas with $\gamma = 4/3$). For $y = \zeta$, Eq. (1.2c) gives $x = (\beta + 7)/2 = 5.5$.

Likewise we have

$y = \zeta/2,\ x = (\beta + 14)/3 = (15\gamma - 13)/3(\gamma - 1) = 6$

$y = \zeta/3,\ x = (\beta + 21)/4 = (11\gamma - 10)/2(\gamma - 1) = 6.25$

All these results indicate the increasing importance of radiation to change the strong shock limit from $x = 4$ to $x = 7$. It should be pointed out that so far our formulation has been a result of neglecting the terms 1 and $\gamma/(\gamma - 1)$ in Eqs. (1.2a) and (1.3a) respectively. In order to validate such an approximation, let us modify the EOS of the γ-law gas as follows. With $\Pi = p_r/p_g$ and $P = p_g + p_r = (1 + \Pi)p_g$, we may write

$$E = \frac{p_g v}{\gamma-1} + 3p_r v = \frac{1+3(\gamma-1)\Pi}{(\gamma-1)(1+\Pi)}Pv = \frac{Pv}{\gamma_H - 1} \tag{4.23}$$

and

$$\gamma_H = \frac{\gamma+4(\gamma-1)\Pi}{1+3(\gamma-1)\Pi} \tag{4.23a}$$

From Eq. (2.1a) we immediately obtain

$$\frac{P_2}{P_1} = \frac{(\gamma_H + 1)v_1 - (\gamma_H - 1)v_2}{(\gamma_H + 1)v_2 - (\gamma_H - 1)v_1} \tag{4.24}$$

and

$$\frac{\rho_2}{\rho_1} = \frac{v_1}{v_2} = \frac{\gamma_H + 1}{\gamma_H - 1} = \frac{7(\gamma - 1)\Pi + (\gamma + 1)}{(\gamma - 1)(1 + \Pi)} \quad \text{as} \ \frac{P_2}{P_1} \to \infty \tag{4.25}$$

Note that γ_H is not a constant because the parameter Π may vary with T^3/ρ. Thus Eq. (4.25) suffices to verify our earlier results exactly. For classical ideal gas ($\Pi \to 0$), Eq. (4.25) gives $\rho_2/\rho_1 = (\gamma + 1)/(\gamma - 1) = 4$ for $\gamma = 5/3$; for photon gas ($\Pi \to \infty$), we obtain $\rho_2/\rho_1 = 7$; and likewise we have

$\Pi = 1, \quad \rho_2/\rho_1 = (4\gamma - 3)/(\gamma - 1) = 5.5$

$\Pi = 2, \quad \rho_2/\rho_1 = (15\gamma - 13)/3(\gamma - 1) = 6$

$\Pi = 3, \quad \rho_2/\rho_1 = (11\gamma - 10)/2(\gamma - 1) = 6.25$

with $\Pi = \zeta/y$ given earlier.

4.6 Hydromagnetic shocks

Magnetohydrodynamics (MHD) is the subject concerning plasma flow in a magnetic field; as electric fields and currents are induced, the magnetic field and electric currents interact and modify each other; and the entire flow is subtly affected along with a variety of hydromagnetic waves. The fundamental principles of MHD are a combination of fluid dynamics and electromagnetic theory. For plasma as a good conductor ($\sigma \to \infty$) without viscosity and thermal diffusion, the basic equations of MHD may be summarized below.

$$\frac{\partial \rho}{\partial t} + \nabla \cdot (\rho \boldsymbol{u}) = 0 \tag{4.26}$$

$$\rho\frac{d\boldsymbol{u}}{dt} = -\nabla p + \mathbf{J} \times \mathbf{B} \tag{4.27}$$

$$\frac{\partial}{\partial t}\left[\rho\left(\varepsilon + \frac{\mathrm{B}^2}{2\mu\rho} + \frac{u^2}{2}\right)\right] + \nabla\cdot\left[\rho\left(\varepsilon + \frac{u^2}{2}\right)\boldsymbol{u} + p\boldsymbol{u} + \mathbf{E} \times \mathbf{H}\right] = 0 \tag{4.28}$$

$$\nabla\cdot\mathbf{B} = 0 \tag{4.29}$$

$$\frac{\partial\mathbf{B}}{\partial t} = -\nabla \times \mathbf{E} \tag{4.30}$$

where **B**, **H**, **E**, and **J** are all vectors denoting the magnetic induction, magnetic field, electric field, and current density respectively. Note that the Lorontz force is explicitly given by

$$\mathbf{J} \times \mathbf{B} = \nabla \times \mathbf{H} \times \mathbf{B} = \nabla\left(\frac{\mathbf{BB}}{\mu} - \frac{\mathbf{B}^2}{2\mu}\right) \tag{4.28a}$$

This is the origin of magnetic tension and pressure. In Eq. (4.28) the Poynting vector is

$$\mathbf{E} \times \mathbf{H} = \mathbf{H} \times (\boldsymbol{u} \times \mathbf{B}) = \frac{\mathbf{B}^2}{\mu}\boldsymbol{u} - (\boldsymbol{u}\cdot\mathbf{B})\frac{\mathbf{B}}{\mu} \tag{4.28a}$$

which denotes the energy flux. Also Ohm's law $\mathbf{J} = \sigma(\mathbf{E} + \boldsymbol{u} \times \mathbf{B})$ is useful in Eqs. (4.30) and (4.28a), viz. $\mathbf{E} = -\boldsymbol{u} \times \mathbf{B}$ as a result of conductivity $\sigma \to \infty$. Magnetic permeability μ provides the relation $\mathbf{B} = \mu\mathbf{H}$. For simplicity the standard MKSA units are used.

All equations of MHD seem simple, but their analysis is often frustrating due to the mathematical implicity of vectors and tensors. Thus a combination of linearized Eqs. (4.26), (4.27), and (4.30) reduces to

$$\frac{\partial^2\boldsymbol{u}'}{\partial t^2} - c_0^2\nabla(\nabla\cdot\boldsymbol{u}') + \boldsymbol{c}_\mathrm{A} \times \nabla \times (\nabla \times \boldsymbol{u}' \times \boldsymbol{c}_\mathrm{A}) = 0 \tag{4.31}$$

with the wave velocity vector $\boldsymbol{c}_A = \mathbf{B_o}/\rho_o\mu$ and $c_o^2 = \partial p/\partial\rho$. Let $\boldsymbol{u}' = \mathbf{A}\exp[i(\mathbf{k}\cdot\mathbf{x} - \omega t)]$

with phase velocity $c = \omega/k$, ω being the angular frequency and k being the wave number. A lengthy manipulation of Eq. (4.31) with its vectors (also θ = angle between vector **B** and u) leads to these equations of dispersion:

$$\omega^2 - k^2 c_A^2 \cos^2\theta = 0 \qquad (4.32a)$$

$$\omega^4 - k^2\left(c_o^2 + c_A^2\right)\omega^2 + k^4 c_o^2 c_A^2 \cos^2\theta = 0 \qquad (4.33a)$$

Accordingly there exist three modes of MHD waves, namely, the fast c_f, intermediate c_i, and slow c_s magnetosonic waves with $c_f \geq c_i \geq c_s$:

$$c_i^2 = c_A^2 \cos^2\theta \qquad (4.32b)$$

$$c_f^2 = \frac{1}{2}\left[\left(c_o^2 + c_A^2\right) + \sqrt{\left(c_o^2 + c_A^2\right)^2 - 4c_o^2 c_A^2 \cos^2\theta}\right] \qquad (4.33b)$$

$$c_s^2 = \frac{1}{2}\left[\left(c_o^2 + c_A^2\right) - \sqrt{\left(c_o^2 + c_A^2\right)^2 - 4c_o^2 c_A^2 \cos^2\theta}\right] \qquad (4.33c)$$

These results are obtainable otherwise by the method of characteristics without linearizing Eqs. (4.26), (4.27), and (4.30) in full scalar forms.

When the amplitudes of the three waves are large in supersonic flow, the c_f and c_s modes steepen to form fast and slow hydromagnetic SW respectively while the c_i mode (Alfvén wave) can propagate without steepening. Note $c_f = c_o$ and $c_i = c_s = c_A$ for $\theta = 0$; $c_f = \sqrt{c_o^2 + c_A^2}$ and $c_i = c_s = 0$ for $\theta = 90°$. Such directional effects prevail also in parallel and perpendicular MHD shocks.

For $\dfrac{\partial}{\partial y} = \dfrac{\partial}{\partial z} = 0$ Eqs. (4.26) - (4.30) lead to the general MHD shock jump conditions in abbreviation $[w] = w_2 - w_1$ as follows.

$$[\rho u_x] = 0 \qquad (4.26a)$$

$$\left[\rho u_x^2 + p + \left(B_y^2 + B_z^2\right)\big/2\mu\right] = 0 \qquad (4.27a)$$

$$\left[\rho u_x u_y - B_x B_y\big/\mu\right] = 0 \qquad (4.27b)$$

$$\left[\rho u_x u_z - B_x B_z/\mu\right] = 0 \tag{4.27c}$$

$$\left[\rho\left(\varepsilon + \frac{u^2}{2}\right)u_x + p u_x + \left(\frac{B_y^2 + B_z^2}{\mu}\right)u_x - \left(u_y B_y + u_z B_z\right)\frac{B_x}{\mu}\right] = 0 \tag{4.28a}$$

$$\left[B_x\right] = 0 \tag{4.29a}$$

$$\left[u_x B_y - u_y B_x\right] = 0 \tag{4.30a}$$

$$\left[u_x B_z - u_z B_x\right] = 0 \tag{4.30b}$$

Now let $u_x = u$ ($u_y = u_z = 0$) and $B_x = B$ ($B_y = B_z = 0$). The above equations reduce to

$$[\rho u] = 0$$

$$[p + \rho u^2] = 0$$

$$\left[h + \frac{u^2}{2}\right] = 0$$

$$[B] = 0$$

only. The first three of these are exactly the same as Eqs. (1.1) - (1.3) in abbreviation. In other words parallel MHD shocks ($\theta = 0$) are the same as ordinary normal SW, the unchanged magnetic field being merely a redundant factor there. On the other hand, perpendicular MHD shocks ($\theta = 90°$) behave with notable magnetic interaction as shown subsequently.

Let $u_x = u$ ($u_y = u_z = 0$) and $B_y = B$ ($B_x = B_z = 0$) for perpendicular MHD shocks ($\theta = 90°$). Then Eqs. (4.26a) - (4.30b) degenerate into only four forms below.

$$[\rho u] = 0$$

$$\left[p + \frac{B^2}{2\mu} + \rho u^2\right] = 0$$

$$\left[h + \frac{B^2}{\mu\rho} + \frac{u^2}{2}\right] = 0$$

$$[Bu] = 0$$

With $p^* = p + \frac{B^2}{2\mu}$ and $h^* = h + \frac{B^2}{\mu\rho}$, the above three equations become the modified forms of Eq. (1.1)- (1.3), and Eq. (1.6) is applicable here as

$$h_2^* - h_1^* = \frac{1}{2}(v_1 + v_2)(p_2^* - p_1^*)$$

For a γ - law gas with $x = \rho_2/\rho_1 = u_1/u_2 = B_2/B_1$, $y = p_2/p_1$, and $\zeta = B_1^2/2\mu p_1$, the above three equation becomes

$$\frac{\gamma}{\gamma - 1}\left(\frac{y}{x} - 1\right) + 2\zeta(x-1) = \frac{1}{2}\left(1 + \frac{1}{x}\right)\left[(y-1) + \zeta(x^2 - 1)\right]$$

Solving this for y, we obtain

$$y = \frac{(\gamma+1)x - (\gamma-1) + \zeta(\gamma-1)(x-1)^3}{(\gamma+1) - (\gamma-1)x}$$

or

$$\frac{p_2}{p_1} = \frac{(\gamma+1)v_1 - (\gamma-1)v_2 + \left(B_1^2/2\mu p_1 v_2^2\right)(\gamma-1)(v_1 - v_2)^3}{(\gamma+1)v_2 - (\gamma-1)v_1} \tag{4.31}$$

This is the shock EOS desired. Note $u_2 = u_1/x$ and $B_2 = xB_1$ (perpendicular MHD shocks with $u_2 < u_1$ and $B_2 > B_1$). For $B_1 = 0$, Eq. (4.31) reduces to Eq. (2.1a) exactly as expected. Note that a perpendicular MHD shock is also a fast shock with $U_f > c_f$. As mentioned above, parameter x indicates B/ρ = constant. Thus we may write

$$\frac{d}{d\rho}\left(\frac{B^2}{2\mu}\right) = \left(\frac{B}{\rho}\right)^2 \frac{d}{d\rho}\left(\frac{\rho^2}{2\mu}\right) = \frac{B^2}{\mu\rho} = c_A^2$$

and hence

$$c^{*2} = dp^*/d\rho = dp/d\rho + \frac{d}{d\rho}\left(\frac{B^2}{2\mu}\right) = c_o^2 + c_A^2 = c_f^2 \text{ for } \theta = 90°.$$

Also it is worth noting that the condition for shock stability is exactly

$$d^2p^*/dv^2 = d^2p/dv^2 + \frac{3}{\mu}\left(\frac{B}{v}\right)^2 > 0.$$

Indeed MHD is a complex subject, but our inquiry has gone far enough to grasp the essential aspects of hydromagnetic shocks. In summary there are at least seven kinds of MHD shocks:

- Parallel shocks ($\mathbf{B}//\boldsymbol{u}$, $\theta = 0$)
- Perpendicular shocks ($\mathbf{B} \perp \boldsymbol{u}$, $\theta = 90°$)
- Slow shocks ($u_1 > c_s$, $0 < \theta < 90°$)
- Fast shocks ($u_1 > c_f$, $0 < \theta \leq 90°$)
- Switch-off shocks ($B_1//u_1$, $\theta \neq 0$, $B_{2y} = 0$)
- Switch-on shocks (fast shock $B_1//u_1$, $B_{2y} > 0$)
- Collisionless shocks with thickness ~ Larmor radius (say, 10^3 km) vs. mean free path of rarefied plasma ~ 10^8 km in outer space (molecular collisions become impossible; parallel collisionless MHD shocks are turbulent; perpendicular collisionless shocks are fast shocks).

4.7 Laser shocks

When a high-power laser beam acts upon a target, the surrounding gas breaks down explosively into plasma which is accompanied by four wave fronts: (1) a laser supported detonation (LSD), (2) a strong SW, (3) a radiation front, and (4) an ionizing front. The LSD may be described by the theory of C-J detonation, and the luminous SW propagates like a spherical blast wave.

Let q denote the laser radiation flux. Then the specific energy input is given by $q/\rho_1 D$. The equations of LSD are

$$\rho_1 u_1 = \rho_2 c_2$$

$$p_1 + \rho_1 u_1^2 = p_2 + \rho_2 c_2^2$$

$$h_1 + \frac{u_1^2}{2} + \frac{q}{\rho_1 D} = h_2 + \frac{c_2^2}{2}$$

with u_1 = D and u_2 = D - c_2 (C-J detonation velocity = D). These equations can be combined with $v = 1/\rho$ to give

$$h_2 - h_1 = \frac{1}{2}(v_1 + v_2)(p_2 - p_1) + v_1 q / D$$

For LSD as a strong shock ($h_2 >> h_1$, $p_2 >> p_1$) and the plasma as a γ-law gas, the above equation reduces to

$$\frac{\gamma p_2 v_2}{\gamma - 1} = \frac{1}{2} p_2(v_1 + v_2) + q_1\left(\frac{v_1 - v_2}{p_2}\right)^{\frac{1}{2}}$$

which gives the shock EOS

$$p_2 = \left[\frac{2q(\gamma-1)(v_1 - v_2)^{1/2}}{(\gamma+1)v_2 - (\gamma-1)v_1}\right]^{2/3} \tag{4.32}$$

Substituting the C-J volume $v_2 = \gamma v_1/(\gamma + 1)$, we obtain

$$p_2 = \left[2q(\gamma^2 - 1)\right]^{2/3} \rho_1^{1/3}/(\gamma + 1) \tag{4.32a}$$

which may be identified with the C-J pressure $p_j = \rho_1 D^2/(\gamma + 1)$. Accordingly we deduce

$$D = \left[2(\gamma^2 - 1)q/\rho_1\right]^{1/3} \tag{4.33}$$

Also we have

$$E_2 = \frac{p_2 v_2}{\gamma - 1} = \frac{\gamma}{\gamma + 1}\left(\frac{4}{\gamma^2 - 1}\right)^{1/3}\left(\frac{q}{\rho_1}\right)^{2/3} \tag{4.34}$$

$$c_2 = \rho_1 D/\rho_2 = \gamma D/(\gamma + 1) = \gamma u_2$$

These results determine the state (p_2, ρ_2, E_2, D, u_2, c_2) of LSD. Note carefully the differences between Eqs. (4.33), (4.34) and (4.12), (4.11b) respectively.

Let the incoming laser beam be focused toward the target with energy yield Y in a solid angle $\Omega = A/4\pi R^2$. Then the energy flux is $q = Y/4\pi\Omega R^2$. Substituting this into Eq. (4.33), we have

$$\frac{dR}{dt} = D = \left[\frac{(\gamma^2 - 1)Y}{2\pi\Omega\rho_1 R^2}\right]^{1/3}$$

which is integrable to give

$$R = \left(\frac{5}{3}\right)^{3/5}\left[\frac{(\gamma^2 - 1)Y}{2\pi\Omega\rho_1}\right]^{1/5} t^{3/5} \tag{4.35}$$

This is the equation of shock trajectory similar to Eq. (2.27). For the luminous SW we have

$$p_H = \frac{\rho_1}{\gamma + 1}\left[\frac{(\gamma^2 - 1)Y}{2\pi\Omega\rho_1 R^2}\right]^{2/3} \tag{4.36}$$

$$p_H R^{4/3} = \frac{\rho_1}{\gamma + 1}\left[\frac{(\gamma^2 - 1)Y}{2\pi\Omega\rho_1}\right]^{2/3} = \text{constant} \tag{4.36a}$$

$$p_H t^{4/3} = \left(\frac{3}{5}\right)^{4/5}\frac{\rho_1}{\gamma + 1}\left[\frac{(\gamma^2 - 1)Y}{2\pi\Omega\rho_1}\right]^{2/5} = \text{constant} \tag{4.36b}$$

These provide the blast wave analogy for the laser shock. See Eqs. (2.28) and (2.29).

Chapter 5 Special Topics

5.1 Shock tubes

Shock waves are ever at large in nature and history, but it is the shock tube that first put SW on hold for measurement and examination. Since the 1940's the development of modern shock tubes has ramified significantly. Today the science and technology of shock tubes can fill many volumes for (i) the generation and testing of SW, (ii) experiments of shock reflection, refraction, and diffraction, (iii) study of supersonic and hypersonic flow, (iv) investigation of radiating, ionizing, and MHD shocks, etc. In what follows we seek to highlight only the theory and applications of shock tubes.

Let us sketch the cconfiguration of a typical shock tube as shown in Fig. 5.1. Note the tube diameter d ~ a few inches, length l ~ 40 to 150 d. At l_4 ~ $l/8$ to $l/5$

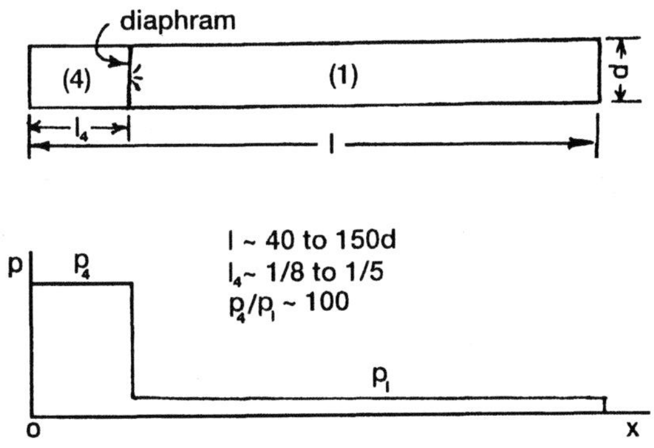

Fig. 5.1 Shock-tube configuration (t = 0).

is a collapsible diaphragm which divides the tube into two chambers. Chamber (4) contains the driver gas at p_4 ~ 10 to 100 atmospheres, and chamber (1) contains the test gas at $p_1 \le 1$ atmosphere. When the diaphragm of separation collapses suddenly, the high-pressure gas rushes into the low-pressure gas, creating three moving fronts **R, C, S** as shown in Fig. 5.2. The shock wave **S** compresses the test gas from state (1) to state (2) which is separated from state (3) by the contact

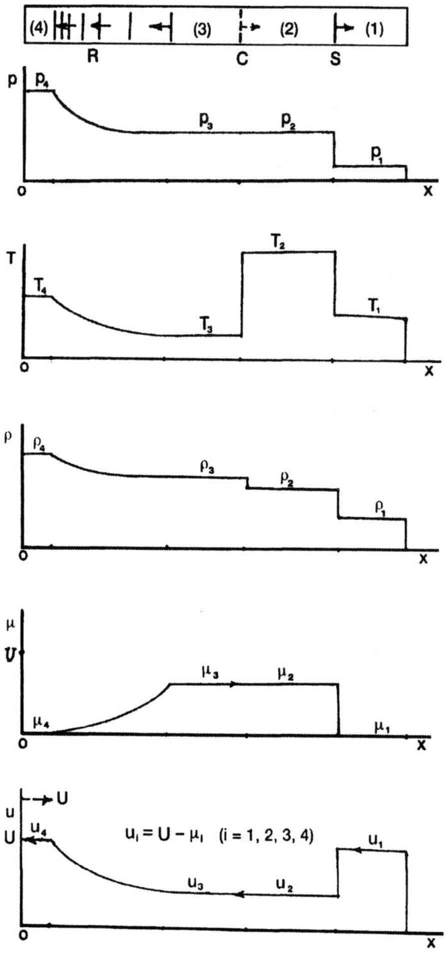

Fig. 5.2 Shock-tube operation (t > 0).

discontinuity **C**. State (3) links to state (4) with a train of rarefaction waves **R**.

Let the initial states of the test and driver gas be given by (p_1, T_1, m_1, γ_1, c_1) and (p_4, T_4, m_4, γ_4, c_4) respectively, where m_1 and m_4 are the molecular weights of the two gases. It is essential to summarize the equations of motion for the three fronts as follows.

For **S** with $\mu_1 = 0$ and $u_1 - u_2 = \mu_2$, Eqs. (2.3), (1.2), and (1.1) give

$$\frac{p_2}{p_1} = 1 + \frac{2\gamma}{\gamma_1 + 1}\left(M_1^2 - 1\right) \tag{5.1a}$$

$$\frac{p_2}{p_1} = 1 + \gamma_1 M_1 \frac{\mu_2}{c_1} \tag{5.1b}$$

$$\mu_2 = \frac{2c_1}{\gamma_1 + 1}\left(M_1 - \frac{1}{M_1}\right) \tag{5.2}$$

For **R**, refer to the Riemann invariant

$$\frac{p_3}{p_4} = \left(\frac{\rho_3}{\rho_4}\right)^{\gamma_4} = \left(\frac{c_3}{c_4}\right)^{\frac{2\gamma_4}{\gamma_4 - 1}} \tag{5.3}$$

$$\mu_3 + \frac{2c_3}{\gamma_4 - 1} = \frac{2c_4}{\gamma_4 - 1} \tag{5.4a}$$

$$\mu_3 = \frac{2c_4}{\gamma_4 - 1}\left[1 - \left(\frac{p_3}{p_4}\right)^{\frac{\gamma_4 - 1}{2\gamma_4}}\right] \tag{5.4b}$$

For **C**, compatibility requires $\mu_2 = \mu_3$ and $p_2 = p_3$ note contact discontinuity: $T_2 \neq T_3$ and $\rho_2 \neq \rho_3$). From Eqs. (5.2) and (5.4b), we obtain

$$\frac{p_3}{p_4} = \left[1 - \frac{c_1}{c_4}\left(\frac{\gamma_4 - 1}{\gamma_1 + 1}\right)\left(M_1 - \frac{1}{M_1}\right)\right]^{\frac{2\gamma_4}{\gamma_4 - 1}} \tag{5.5}$$

From this and Eq. (5.1a), we may write

$$\frac{p_4}{p_1} = \frac{p_2/p_1}{p_3/p_4} = \left[1 + \frac{2\gamma_1}{\gamma_1 + 1}\left(M_1^2 - 1\right)\right]\left[1 - \frac{c_1}{c_4}\left(\frac{\gamma_4 - 1}{\gamma_1 + 1}\right)\left(M_1 - \frac{1}{M_1}\right)\right]^{-\frac{2\gamma_4}{\gamma_4 - 1}} \tag{5.6}$$

It should be pointed out that the shock Mach number M_1 is the key to the theory of shock tubes. Now the denominator on the right-hand side of Eq. (5.6) must be zero as $p_4/p_1 \to \infty$. Accordingly we deduce

$$M_1 = \left(\frac{\gamma_1+1}{\gamma_4-1}\right)\frac{c_4}{c_1} = \frac{\gamma_1+1}{\gamma_4-1}\left(\frac{\gamma_4 T_4 m_1}{\gamma_1 T_1 m_4}\right)^{1/2} \tag{5.7a}$$

$$\frac{p_2}{p_1} = \frac{2\gamma_1}{\gamma_1+1}M_1^2 = \frac{2\gamma_4(\gamma_1+1)}{(\gamma_4-1)^2}\left(\frac{T_4 m_1}{T_1 m_4}\right) \tag{5.7b}$$

for a strong shock **S** ($M_1^2 >> 1$). The significance of Eqs. (5.7a) and (5.7b) is now illustrated by the numerical results below. The ratio c_4/c_1 is given by $(\gamma_4 m_1/\gamma_1 m_4)^{1/2}$, assuming

test gas	m_1	γ_1	driver gas	m_4	γ_4	c_4/c_1	M_1	p_2/p_1
air	28.9	4/3	air	28.9	4/3	1	7	56
argon	40	5/3	helium	4	5/3	3.16	12.6	198
argon	40	5/3	hydrogen	2	7/5	4.10	27.3	932
xenon	131	5/3	hydrogen	2	7/5	7.42	49.5	3063

$T_4 = T_1$ initially. These results show the combination advantages of light driver-gas and heavy test-gas. Note that the larger is m_1/m_4, the higher is c_4/c_1 for a strong shock **S**. If we raise T_4 above T_1, the resultant shock **S** will be stronger. This is accomplished by heating the driver gas electrically or thermally. The so-called combustion shock tube is to synchronize the diaphragm breakdown and the ignition of a driver gas mixture (e.g. 8% O_2 + 92% H_2 or 10% O_2 + 20% H_2 + 70% He) so that the mixture burns at a constant pressure for the generation of shock **S**. All these considerations are conceptually sound, and Eqs. (5.7a) and (5.7b) can serve as an engineering guide to the design of high-performance shock tubes. In fact, several other innovations are practical for the production of strong shock **S** as well. Here we mention: (i) double-diaphragm shock tube, (ii) area-change shock tube with $A_4/A_1 > 1$, (iii) hypersonic shock tunnel with a divergent nozzle attached to the test-gas chamber, (iv) magnetically driven shock tube with a perpendicular MHD shock in deuterium to reach $M_1 \sim 250$.

Both conventional and advanced shock tubes are useful for the study of shock waves, especially in search of shock reflection (regular & Mach reflections), refraction, and diffraction. Shock tubes are an indispensable tool to examine the gas kinetic degrees of freedom, dissociation, ionization, and chemical reaction due to shock heating. The exploration of negative and phase-transition SW was made possible also by shock-tube experiments. In aerodynamic laboratory shock tube can serve to simulate the flight environment for sweptback wings and missiles. Due to the complexity of theory, shock-tube experiments are the primary steps to take up the study of shock focusing, imploding SW, and MHD shocks. To sum up, shock tubes have contributed immensely to the experimentation as well as the theorization of SW.

5.2 Hypervelocity impact

The specialized study of impact deals with the mechanics of stress waves, deformations, and/or breakages. The damaging effects of impact are especially of importance in modern ballistics and weaponry. It is convenient to investigate impact phenomena according to the five categories:

- Elastic impact $0 < V < V_1$, $\zeta < 10^{-5}$ ($V \sim 1$ m/sec)
- Plastic impact $V_1 < V < V_2$, $10^{-5} < \zeta < 1$ ($V \sim 5$ to 500 m/sec)
- Ballistic impact $V_2 < V < V_3$, $1 < \zeta < 100$ (V up to 3 km/sec)
- Hypervelocity impact $V_3 < V < 3V_3$, $\zeta > 10^3$ ($V \sim 5$ km/sec)
- Explosive impact $V > 3V_3$, $\zeta > 10^4$ ($V \sim 15$ km/sec)

with V = striking velocity, threshold $V_1 = V_2^2\left(c_b^{-1} + c_l^{-1}\right)$, $V_2^2 = \sigma_y/\rho$, $V_3^2 = B/\rho$ (B = bulk modulus of compression, $c_b^2 = E/\rho$, $c_l^2 = (\lambda + 2\mu)/\rho$, E = Young's modulus, λ, μ = Lamé constants, ρ = density), and damage number $\zeta = \rho V^2/\sigma_y$. We will leave out the first three categories which belong to the study of elasticity and plasticity, seismology, and penetration ballistics. Here it is of much relevance to focus our inquiry on the latter categories regarding solid- or liquid-to-solid impact. Only computer simulation is feasible to analyze the phenomena of solid- or liquid-to-liquid impact (say, the splash of a waterdrop treated elsewhere).

Let us now clarify the concept of hypervelocity (HV) further. From Eq. (2.6a) we find out T ~ 1500°K for a strong SW with $M_1 \sim 4$ in a monatomic gas. From Eq. (4.10) we obtain $p_j > 100$ kb for C-J detonation to induce a SW in a solid by contact explosion ($\gamma \sim 3$, $D \sim 2$ km/sec, $\rho > 1$ g/cm^3). Note that TNT has a specific energy yield $Q \sim 10^3$ cal/g = 4.2×10^{10} erg/g or (cm/sec)^2 and an energy density $\rho Q \sim 10^{11}$ erg/cm^3 or (dyne/cm^2). These data and units appear to suggest the shock intensity in solids to be $p_H > 100$ kb and U > 1 km/sec. In view of $p_H \sim \rho U^2$ and bulk modulus $B = \rho c^2$, we have $M^2 = U^2/c^2 \sim p/B > 1$ consistently, too. Earlier we learned that strong SW can be generated in a shock tube with light gas (e.g. H_2 or He) as the driver. Likewise, light gas gun ($l > 100$ d) has become a standard tool to produce SW in solids. This is a modified shock tube with the diaphragm substituted by a disk projectile for high muzzle velocity to inflict the desired impact. Thus the specific kinetic energy of a projectile at velocity above 1 km/sec is sufficient to produce SW as well as by 1 gram of TNT. Such an equivalence of the impact effect to the explosion effect marks out the threshold of HV clearly. HV impact is of practical importance not only in the study of SW and EOS but also in terminal ballistics and space flight. Moreover, explosive impact is essentially an advanced HV impact which generates strong SW and crater (e.g. the spectacular impact of a giant meteorite on earth).

At the instant of HV impact, two SW are propagating separately in the striker and in the barricade, and the interface becomes a contact discontinuity with $u_t = u$, $u_j = V - u$, and $p_t = p_j$ (here subscript t referring to target and j to projectile or jet later). Now with Eq. (3.32a), it is expedient to formulate the basic equations of HV impact as follows.

$$p_t = \rho_t u_t U_t = \rho_t u(c_t + s_t u)$$

$$p_j = \rho_j u_j U_j = \rho_j (V - u)[c_j + s_j (V - u)]$$

Let $r = \rho_t/\rho_j$ and equate the above two equations. The result is then

$$Au^2 - Bu + C = 0 \tag{5.8}$$

$$A = s_j - rs_t$$

$$B = c_j + rc_t + 2s_jV$$

$$C = \left(c_j + s_jV\right)V$$

Once u is determined from Eq. (5.8), a complete solution is feasible for the HV impact. The following two special cases are worth noting.

For the impactor and target to be made of the same material, we have $r = 1$ ($\rho_t = \rho_j = \rho_o$), $c_t = c_j = c_o$, $s_j = s_t = s$, and hence

$$A = 0$$

$$B = 2(c_o + sV)$$

$$C = (c_o + sV)V$$

Substituting these into Eq. (5.8), we obtain

$$u = C/B = V/2 \tag{5.8a}$$

$$U = c_o + sV/2 = U_j = U_t$$

$$p_H = \rho_o uU = \frac{1}{2}\rho_o V^2\left(\frac{c_o}{V} + \frac{s}{2}\right) = p_j = p_t$$

$$M = 1 + sV/2c_o$$

From these and $s = 3/2$ ($s = (\Gamma_o + 1)/2$ for Grüneisen $\Gamma_o \sim 2$), we note

$$V = c_o,\ M = 1.75,\ p_H = \frac{7}{8}\rho_o V^2 = \frac{7}{8}\rho_o c_o^2 = \frac{7}{8}B_o$$

$$V = 2c_o,\ M = 2.5,\ p_H = \frac{5}{8}\rho_o V^2 = \frac{5}{2}\rho_o c_o^2 = \frac{5}{2}B_o$$

$$V = 3c_o, \quad M = 3.25, \quad p_H = \frac{13}{24}\rho_o V^2 = \frac{39}{8}\rho_o c_o^2 = \frac{39}{8}B_o$$

For unlike-solid impact, we should have (a) $u_t < V/2$ for hard target and (b) $u_t > V/2$ for soft target as a result of shock-impedance mismatch (see Fig. 5.3).

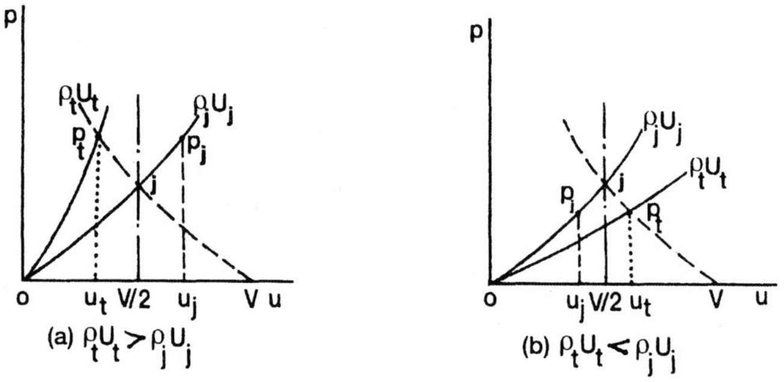

Fig. 5.3 Mirror-image approximation of projectile Hugoniot for shock-impedance mismatch during HV impact: (a) hard target, (b) soft target.

Note that curve Vjp_t is the mirror image of curve ojp_j (similar for Vp_tj and op_jj). Apparently, soft material (e.g. plastic foam) can serve as a damper of impact shock.

For a slender projectile to impact on a thick target at HV > 10 km/sec, we may write $s_j V >> c_j, c_t; s_j \approx s_t = s$, and hence

$$A = (1-r)s$$
$$B = 2sV$$
$$C = sV^2$$
$$(1-r)su^2 - 2sVu + sV^2 = 0 \tag{5.8b}$$

Apparently Eq. (5.8b) has a root

$$u_t = \frac{2V - \sqrt{(2V)^2 - 4(1-r)V^2}}{2(1-r)} = \frac{1-\sqrt{r}}{1-r}V = \frac{V}{1+\sqrt{\dfrac{\rho_t}{\rho_j}}} \tag{5.8c}$$

which is the equation of shaped-charge jet penetration to be proved otherwise subsequently. The impact pressure is

$$p_{\mathrm{H}} = s\rho_t u^2 = \rho_j \mathrm{V}^2\left(\frac{sr}{1+r+2\sqrt{r}}\right) \tag{5.8d}$$

Besides the generation and propagation of SW, HV impact also causes damages in both the projectile and the target, which are the main theme of terminal ballistics. Here we seek to illustrate this effect of impact by a simplified theory. Let the penetration depth be X when the projectile with an impact velocity V to crater the target is stopped. Then we may write the equation of dynamic balance:

$$m_j \frac{du}{dt} = -f = -bx^2$$

which is integrable as

$$\int_{\mathrm{V}}^{\mathrm{o}} m_j u du = -\int_{\mathrm{o}}^{\mathrm{X}} bx^2 dx$$

or

$$\frac{1}{2} m_j \mathrm{V}^2 = \frac{1}{3} b\mathrm{X}^3 \tag{5.9a}$$

Note that f is the target resistance with coefficient b depending on material hardness ($\mathrm{H_t}$) and the crater shape. Thus Eq. (5.9a) provides a rule of thumb for HV cratering, viz. the crater volume ($m_t/\rho_t \propto \mathrm{X}^3$) being proportional to the impact energy. This rule may be expressed nondimensionally as the cratering efficiency

$$\eta = \frac{(m_t/\rho_t)\mathrm{H_t}}{m_j \mathrm{V}^2/2} = \frac{2m_t/m_j}{\rho_t \mathrm{V}^2/\mathrm{H_t}} \propto \frac{\rho_t \mathrm{X}^3/\rho_j d^3}{\rho_t \mathrm{V}^2/\mathrm{H_t}} \tag{5.9b}$$

where d is the projectile caliber and $\rho_t\mathrm{V}^2/\mathrm{H_t}$ is a parameter similar to the damage number ζ. On the other hand, the cratering efficiency is strongly dependent on the projectile and target materials: $\eta \propto \rho_j/\rho_t$. Accordingly, we deduce

$$\frac{X}{d} = k\left(\frac{\rho_j}{\rho_t}\right)^{\frac{2}{3}}\left(\frac{\rho_t V^2}{H_t}\right)^{\frac{1}{3}} \tag{5.9c}$$

where the empirical constant k is to be determined by data fitting.

5.3 Shaped-charge jet penetration

A shaped charge is a high-explosive warhead designed for defeating the armor as sketched in Fig. 5.4. Note d = weapon caliber, s = stand-off for more penetrating power, l = charge length for high-speed jetting, t = armor thickness, and θ = cone

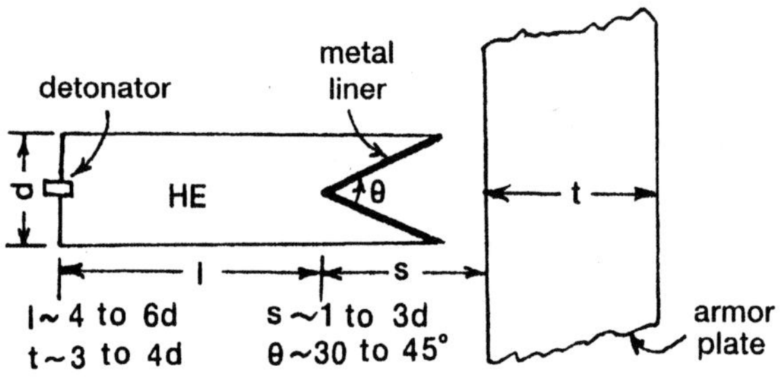

Fig. 5.4 Shaped-charge configuration.

angle of metal liner. When a detonation wave intercepts the metal liner, the latter will be crushed into a liquid jet which pierces the armor. The jet penetration is depicted in Fig. 5.5(a) with jet velocity, V > u, the rate of penetration. Fig. 5.5(b) shows the material flow similar to two colliding jets (A_10, A_20) with two resultant streams (oe_1, oe_2). According to the Bernoulli equation, we have

$$p_o + \frac{1}{2}\rho_j(u - V)^2 = p_o + \frac{1}{2}\rho_t u^2$$

and hence

$$u = V \Big/ \left(1 + \sqrt{\frac{\rho_t}{\rho_j}} \right) \qquad (5.10a)$$

which is the same result as Eq. (5.8c). Assuming that the entire jet l_j is consumed to

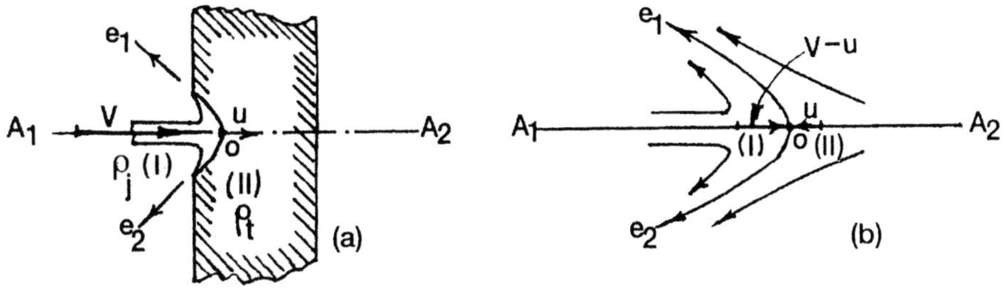

Fig. 5.5 Hydrodynamics of shaped-charge jet penetration: (a) jet piercing armor, (b) analogy of two colliding jets.

penetrate a depth X in time T, we may write

$$l_j = (V - u)T$$

$$X = uT = ul_j/(V - u) = l_j \sqrt{\frac{\rho_j}{\rho_t}} \qquad (5.10b)$$

From Eq. (5.10b) it is inferred that the harder the jet, the deeper the penetration. Thus far we have only considered the elementary theory of shaped-charge jet penetration, and it ought to be further modified for accuracy.

It may be of relevant interest to mention the performance of shaped-charge jet penetration in the anti-tank (AT) warfare. Two types of AT weapons have been used: (a) high-explosive-antitank (HEAT) shells, grenades, and mines; (b) high-explosive-plastic projectile (HEP). The U.S. Army bazooka is a rocket-propelled HEAT shell, and the German Panzerfaust is a HEAT grenade. Since shaped-charge jet can never penetrate armor thicker than 10 inches, many techniques prevail to protect tanks by using heavy or composite armors, spikes to spoil the stand-off of HEAT round, detonation of small shaped charges against the attacking jet, coating of oxidizing material to dispose the incoming jet, etc. As a result, HEP uses a new principle to disable tanks. Because of its plastic filler (comp C-4 or A-3), this round tends to spread over the target surface on impact. Detonation of the explosive charge sets up SW which cause spalling of the tank interior surface without necessarily penetrating the exterior surface. HEP are also effective against concrete, timber barricades and bunkers.

5.4 Inertial confinement fusion (ICF) by imploding SW

For decades of research it has not been able to achieve a controllable nuclear fusion in the tokamak (a Russian word for magnetic confinement). On the other hand it is worth noting that the world's nuclear club countries have tested powerful H-bombs. These use solid lithium deuteride (LiD) as the source of deuterium-titrium (DT) fuel, and the thermonuclear reactions are triggered by the heat and neutrons from their surrogate A-bombs. Note $kT = \varepsilon = 1ev = 1.60 \times 10^{-12}$ erg and $T = \varepsilon/k =$

$$\left.\begin{array}{l} {}^{6}_{3}\mathrm{Li}_3 + {}^{1}_{0}n_1 \longrightarrow {}^{4}_{2}\mathrm{He}_2 + {}^{3}_{1}\mathrm{H}_2 + 5Mev \\ {}^{2}_{1}\mathrm{H}_1 + {}^{3}_{1}\mathrm{H}_2 \xrightarrow{10^8\mathrm{K}} {}^{4}_{2}\mathrm{He}_2 + {}^{1}_{0}n_1 + 17.6Mev \end{array}\right\} \Leftarrow \text{A - bomb}$$

(+

$${}^{6}_{3}\mathrm{Li}_3 + {}^{2}_{1}\mathrm{H}_1 \longrightarrow 2\left({}^{4}_{2}\mathrm{He}_2\right) + 22.6Mev \ (\rightarrow \text{explosion energy H - bomb})$$

$(1.60 \times 10^{-12}) \div (1.38 \times 10^{-16}) \sim 10^4$K. Thus the center of a H-bomb is subjected to $p \sim 10^3$Mb (or 10^9 atmosphere) and $T \sim 10^8$K (cf. Sun's center $p \sim$ 2,500Mb, $T \sim 1.6 \times 10^7$°K; supernova $T \sim 3.5 \times 10^9$K). Since the explosion of H-bomb is an unconfined thermonuclear fusion of DT, an innovation of ICF began with laser-induced imploding SW to do the job during the 1960's. Today it is promising to anticipate a fruitful ICF of DT by SW.

In the ICF experiment a small DT-fuel ball is covered with a layer of ablative material. When high-power laser pulses act upon the fuel ball, ablative plasma blows off like a rocket exhaust to induce converging SW. As successive converging

SW arrive at the center simultaneously, the compression is strong enough to ignite the DT core. Then a supersonic heat wave spreads out to burn the remaining fuel with ample energy yield. The scheme of thermonuclear fusion is shown below.

$$\left.\begin{aligned}
&{}^2_1H_1 + {}^2_1H_1 \xrightarrow{2\times10^8 K} {}^3_2He_1 + {}^1_0n_1 + 3.27Mev \\
&{}^2_1H_1 + {}^2_1H_1 \xrightarrow{5\times10^8 K} {}^3_1H_2 + {}^1_1p_0 + 4.03Mev \\
&{}^2_1H_1 + {}^3_1H_2 \xrightarrow{5\times10^7 K} {}^4_2He_2 + {}^1_0n_1 + 17.6Mev
\end{aligned}\right\} \Leftarrow \text{laser - induced imploding SW}$$

$$\begin{aligned}
&{}^2_1H_1 + {}^3_2He_1 \xrightarrow{5\times10^8 K} {}^4_2He_2 + {}^1_1p_0 + 18.3Mev \qquad +) \\
\hline
&6\left({}^2_1H_1\right) \xrightarrow{10^8 K} 2\left({}^4_2H_2\right) + 2\left({}^1_1p_0\right) + 2\left({}^1_0n_1\right) + 43.2Mev \quad \text{(ICF energy)}
\end{aligned}$$

Now it is especially of interest to look into the shock compression of DT fuel. According to Lawson's criterion for nuclear fusion: $\rho r > 3$ [g/cm^2], we must have

$$\frac{\rho_2 r_2}{\rho_1 r_1} = \frac{r_2\left(m/r_2^3\right)}{r_1\left(m/r_1^3\right)} = \left(\frac{r_1}{r_2}\right)^2 \sim 150, \quad \frac{r_1}{r_2} \sim 12$$

and hence

$$\frac{\rho_2}{\rho_1} = \left(\frac{r_1}{r_2}\right)^3 \sim 2000$$

for liquid DT density $\rho_1 = 0.2$ [g/cm3] and the ball radius $r_1 = 0.1$ cm. As indicated earlier in H-bomb, T_2 must be of the order of 10^8°K. These requirements can be met by imploding shock compression. In Section 2.4 we mentioned $\rho_2/\rho_1 = 23$ as attainable by a single imploding spherical shock. So another innovation is to approximate the successive shock compression by an isentropic compression. Let us consider n plane SW proceeding successively with equal steps:

$$\frac{\rho_2}{\rho_1} = \frac{\rho_3}{\rho_2} = \ldots = \frac{\rho_n}{\rho_{n-1}} = \zeta \quad \therefore \quad \frac{\rho_n}{\rho_1} = \zeta^{n-1} \tag{5.11a}$$

$$\frac{p_2}{p_1} = \frac{p_3}{p_2} = \ldots = \frac{p_n}{p_{n-1}} = 1 + \Delta, \quad \frac{p_n}{p_1} = (1+\Delta)^{n-1} \tag{5.11b}$$

From Eq. (2.1a) we may write

$$\frac{p_i}{p_{i-1}} = \frac{(\gamma+1)\zeta-(\gamma-1)}{(\gamma+1)-(\gamma-1)\zeta} \quad \text{and} \quad \Delta = \frac{p_i}{p_{i-1}} - 1 = \frac{2\gamma(\zeta-1)}{(\gamma+1)-(\gamma-1)\zeta} \tag{5.11c}$$

The timing of these SW is such that they all arrive at time t at the same place of high compression, without overtaking each other. Thus we have

$$\mathrm{U_n}(t-t_n) \le \mathrm{U_{n\text{-}1}}(t-t_{n-1}) \le \ldots \le \mathrm{U_1}t$$

and hence

$$\frac{\mathrm{U_n}}{\mathrm{U_1}} \le \frac{t}{t-t_n} \tag{5.11d}$$

For an isentrope we may write

$$\frac{p_n}{p_1} = \left(\frac{\rho_n}{\rho_1}\right)^{\gamma_e} \tag{5.12a}$$

$$\frac{\mathrm{T_n}}{\mathrm{T_1}} = \left(\frac{\rho_n}{\rho_1}\right)^{\gamma_e-1} \tag{5.12b}$$

with γ_e to be determined as follows. Substituting Eqs. (5.11a) and (5.11b) into Eq. (5.12a), we deduce

$$\gamma_e = \log(1+\Delta)/\log\zeta \tag{5.13}$$

where ζ and Δ are related by Eq. (5.11c). Here we will illustrate a shortcut with $\rho_n/\rho_1 = 10^3$ and $\mathrm{T_n}/\mathrm{T_1} = 10^6$ for $\mathrm{T_1} \sim 300°\mathrm{K}$ (see earlier requirements). From these and Eq. (5.12b), we obtain $10^6 = 10^{3(\gamma_e-1)}$ and hence $\gamma_e = 3$. Substituting this in Eq. (5.12a), we deduce $p_n/p_1 = (10^3)^{\gamma_e} = 10^9$. In other words, the final shock pressure $p_n = 10^9$ bars = 1,000 Mb for p_1 = 1 bar (or 1 atmosphere). Now Eqs. (5.11a) and (5.13) can serve to determine n and Δ. Since the strong shock limit requires $\zeta = (\gamma+1)/(\gamma-1) = 4$, we may choose ζ = 3. Thus we deduce $n = 1 + 3/\log 3 \approx 7$ and $\Delta = 3^3 - 1 = 26$. For $\zeta = 2$ we must have $n = 1 + 3/\log 2 \approx 11$ and $\Delta = 7$. All these results are consistent with our earlier assessment. However, it should be noted that the ICF research has many obstacles and new developments. Since the mid 1970's the so-called indirect-drive fusion has received more attention than the direct-drive fusion by laser-induced implosion as described herewith. For more details, see Reference 27 regarding the direct-drive fusion, but Reference 28 concerns mainly with the indirect-drive fusion.

5.5 Cosmic and isothermal shocks

In outer space there are various kinds of SW, such as the bow shocks (collisionless plasma shocks) which confront all planets around the sun. These SW are a result of the interaction of solar winds with the many planetary magnetospheres. Solar flares and sunspot activities also send out plasma SW. Afar to the distant galaxy, cosmic shocks are commonly associated with the evolution of stars. Here we seek to describe two simplified models based on the classical ideal gas (H_2, He) EOS.

Supernova is the sudden explosion of a massive star ($M > 8\ M_0$), ejecting half its mass with a velocity $c/100$ (c = speed of light) and kinetic energy $E_0 \sim 10^{51}$ ergs (number density $n_0 \sim 10^6$ particles/cm^3). Let us adopt the point-explosion theory of strong spherical shocks with density jump $\rho/\rho_o = (\gamma+1)/(\gamma-1) = 4$. $\varepsilon = \dfrac{p/\rho}{\gamma-1} = \dfrac{2U^2}{(\gamma+1)^2} = \dfrac{9}{32}U^2$. $\varepsilon_k = \dfrac{u^2}{2} = \dfrac{1}{2}\left(1-\dfrac{\rho_o}{\rho}\right)^2 U^2 = \dfrac{9}{32}U^2$. From these we obtain

$$E_o = \frac{4}{3}\pi R^3 \rho_o (\varepsilon + \varepsilon_k) = \frac{3}{4}\pi\rho_o R^3 \dot{R}^2 \tag{5.14a}$$

with $U = \dot{R} = dR/dt$ = shock velocity. Apparently Eq. (5.14a) is integrable to give

$$R = \left(\frac{25E_o}{3\pi\rho_o}\right)^{1/5} t^{2/5} \tag{5.14b}$$

$$U = \dot{R} = \frac{2R}{5t} \tag{5.14c}$$

From these we deduce the shock pressure

$$p_H = \frac{2\rho_o U^2}{\gamma+1} = \frac{3}{4}\rho_o U^2 = \frac{3\rho_o R^2}{25t^2} = \left(\frac{3\rho_o}{25}\right)^{3/5}\left(\frac{E_o}{\pi}\right)^{2/5} t^{-6/5} \tag{5.15a}$$

and hence

$$p_H t^{6/5} = \left(\frac{3\rho_o}{25}\right)^{3/5}\left(\frac{E_o}{\pi}\right)^{2/5} = \text{constant} \tag{5.15b}$$

where the constants are given by substituting $\rho_o = n_o m_H$ (m_H = mass of hydrogen atom) and E_o as mentioned earlier. For strong shock limit see Eqs. (2.3b) and (2.5a). Also compare the above results with the modeling of spherical blast waves in Section 2.3.

Luminous hot stars are losing masses ($\dot{M}_* = 10^{-6}\, M_o/yr$) which are referred to as the stellar wind with velocity $V_* = 2{,}000$ km/sec and energy $\dot{E}_* = \frac{1}{2}\dot{M}_* V_*^2 \sim 10^{36}$ erg/sec (Note $M_o = 1.99 \times 10^{30}$ kg = sun's mass). Also they are fully ionized gas (H^+, e^-), emitting radiation with an ionization front (I-F). The expansion of I-F and the flow of stellar wind are more or less similar to the events in a shock tube (See Fig. 5.2). As shown in Fig. 5.6 four regions are noticeable in the whole process: (a) hot star

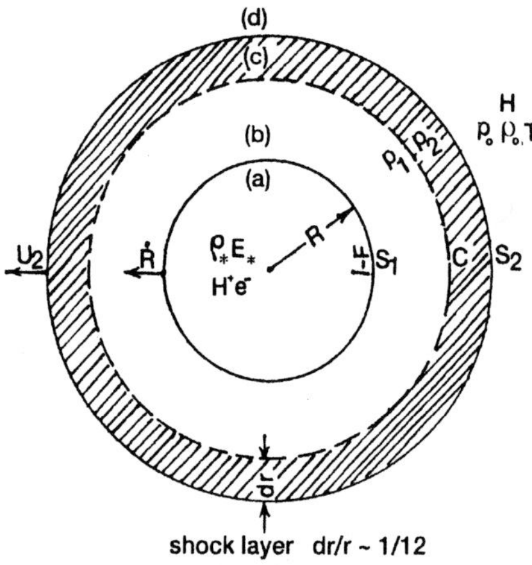

Fig. 5.6 Formation and propagation of ionization front (I-F) and isothermal shock (S_2). Region (a) hot star generating I-F and S_1, (b) plasma shock S_1 and contact discontinuity C being strongly radiant, (c) shock layer absorbing radiation and defining S_2, and (d) undisturbed gas.

generating I-F, (b) stellar wind driving S_1 which propagates together with I-F, (c) contact discontinuity C separating two shocked states with $p_1 = p_2$, and (d) isothermal shock S_2 advancing into the undisturbed gas.

The equations of motion and energy for S_1 and I-F may be given simply by

$$\frac{d}{dt}\left(\frac{4}{3}\pi R^3 \rho_* \dot{R}\right) = 4\pi R^2 p_1 \tag{5.16a}$$

$$p_1 = \rho_*\left(\dot{R}^2 + \frac{1}{3}R\ddot{R}\right) \tag{5.16b}$$

$$\dot{E}_* = \frac{d}{dt}\left(\frac{4}{3}\pi R^3 \frac{p_1}{\gamma - 1}\right) + p_1 \frac{d}{dt}\left(\frac{4}{3}\pi R^3\right) \tag{5.17a}$$

$$R^4\dddot{R} + 12R^3\dot{R}\ddot{R} + 15R^2\dot{R}^3 = 3\dot{E}_*/2\pi\rho_* \tag{5.17b}$$

where Eq. (5.17b) is a direct result of combining Eqs. (5.16b) and (5.17a) with $\gamma = 5/3$. As the spherical flow is self similar, we may let $R = At^{\alpha}$ in Eq. (5.17b) which gives

$$A^5\left[\alpha(\alpha - 1)(\alpha - 2) + 12\alpha^2(\alpha - 1) + 15\alpha^3\right]t^{5\alpha - 3} = 3\dot{E}_*/2\pi\rho_* \tag{5.17c}$$

Since the right-hand side of Eq. (5.17c) is given by a constant we must have $5\alpha - 3 = 0$ to maintain the constancy. With $\alpha = 5/3$, we deduce

$$A = \left(\frac{125}{154\pi}\right)^{\frac{1}{5}}\left(\frac{\dot{E}_*}{\rho_*}\right)^{\frac{1}{5}} = \text{constant} \tag{5.18a}$$

$$R = \left(\frac{125}{154\pi}\right)^{\frac{1}{5}}\left(\frac{\dot{E}_*}{\rho_*}\right)^{\frac{1}{5}} t^{3/5} \tag{5.18b}$$

$$U_1 = \dot{R} = \frac{3R}{5t} \tag{5.18c}$$

$$\dot{U}_1 = \ddot{R} = -\frac{6R}{25t^2} \tag{5.18d}$$

$$p_1 = \rho_*\left(\dot{R}^2 + R\ddot{R}/3\right) = \frac{7}{9}\rho_* U_1^2 = \frac{7\rho_* R^2}{25t^2} \qquad (5.19a)$$

and hence

$$p_1 t^{4/5} = \left(\frac{7\rho_*}{25}\right)^{3/5}\left(\frac{5\dot{E}_*}{22\pi}\right)^{2/5} = \text{constant} \qquad (5.19b)$$

As Eqs. (5.18d) and (5.18c) imply deceleration, The I-F will move subsonically as a rarefaction alone. Compare the final results, Eqs. (5.18b), (5.18c), (5.19a) and (5.19b) with Eqs. (5.14b), (5.14c), (5.15a), and (5.15b) respectively. In the early stage, both cosmic shocks are approximated closely by the strong shock limit.

Now it is interesting to consider the isothermal shock S_2. Since radiant heat conduction is important at high temperature and since the mean free path of photons is larger than the shock thickness *dr* (Fig 5.6), shock heating is balanced by radiation cooling and hence a temperature jump is not possible ($T_2 = T_0$ across the shock front S_2). The following equations govern the isothermal shock jump:

$$\rho_o u_o = \rho_2 u_2$$

$$p_o + \rho_o u_o^2 = p_2 + \rho_2 u_2^2$$

$$c_o^2 = \frac{p_o}{\rho_o} = \frac{p_2}{\rho_2} \quad \text{(isothermal EOS or Boyle's law)}$$

Combining these equations for $p_2 >> p_0$ (strong shock), we may write

$$\rho_2 c_o^2 = p_2 \approx \rho_o u_o^2 - \rho_2 u_2^2 = \rho_2 u_2 u_o - \rho_2 u_2^2$$

and hence

$$u_2^2 - u_o u_2 + c_o^2 = 0 \qquad (5.20a)$$

$$u_2 = \frac{u_o}{2}\left(1 \pm \sqrt{1 - \frac{4c_o^2}{u_o^2}}\right) \approx \frac{u_o}{2}\left[1 \pm \left(1 - \frac{2c_o^2}{u_o^2}\right)\right]$$

Apparently the + sign leads to a trivial solution ($u_2 \approx u_o$) and a meaningful solution is

$$u_2 = c_o^2/u_o = c_o/M_o \tag{5.20b}$$

$$p_2 = \rho_2 c_o^2 = \rho_2 u_2 u_o = \rho_o u_o^2 = \rho_o U_2^2 \tag{5.20c}$$

$$\rho_2/\rho_o = u_o/u_2 = u_o^2/c_o^2 = M_o^2 \tag{5.20d}$$

where the shock Mach number is $M_o = u_o/c_o$ with $u_o = U_2$ = shock velocity in gas at rest. Again this solution approximates the strong shock limit with $\gamma = 1$ [cf. Eqs. (2.3b), (5.15a), (5.19a), and (5.20c)]. From Eqs. (2.5a) and (5.20d) we have $\gamma = (M_o^2+1)/(M_o^2-1)$ =1, too. It should be remarked that an isothermal shock prevails also in a nuclear explosion because the fireball is an isothermal sphere.

5.6 Shock focusing and extracorporeal shock wave lithotripsy (ESWL)

Noticeably, converging shocks (implosion) and shaped-charge effects are all to gain shock strength sharply. Thus the idea of shock focusing has aroused the researcher's imagination wildly. Experimental study indicates that shock reflection from a concave wall turns out to converge at a focus with the result of a pressure spike. This is the gasdynamic focus, and complex wave patterns occur near it due to Mach reflection and shock diffraction. Numerical simulation of shock focusing verifies the observed results satisfactorily. From the linear-wave theory, light and acoustic rays can be brought to a focus by lenses or mirrors. This is the geometric focus. Note that the gasdynamic focus may not coincide with the geometric focus. The departure is due to the complexity of nonlinear shock propagation and interaction. Hemispherical, parabolic, elliptic and other concave reflectors have been tested for shock focusing in air, water, or Plexiglass (PMMA). Here it is of special interest to illustrate the focusing of underwater SW by an ellipsoidal bath for the pulverization of kidney stones. The pertinent medical procedure is known as extracorporeal shock wave lithotripsy (ESWL) which is surgery-free, safe, and less costly.

In 1981 Dornier Med Tech of Germany was the first to introduce its commercial lithotripter for successful clinic operation. Fig. 5.7 shows the technical concepts and procedure layout obviously. This prototype lithotripter is still used in some German hospitals, but newer designs replace the water bath with a revolvable table-

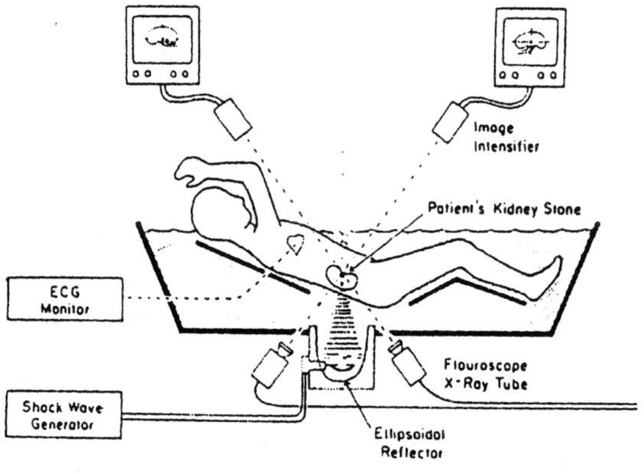

Fig. 5.7 Prototype lithotripter and its clinic layout.

top (see Fig. 5.8). In some lithotripters the therapy unit uses a spherical bowl as the reflector to focus the shock pulses, and in others the SW are focused with an

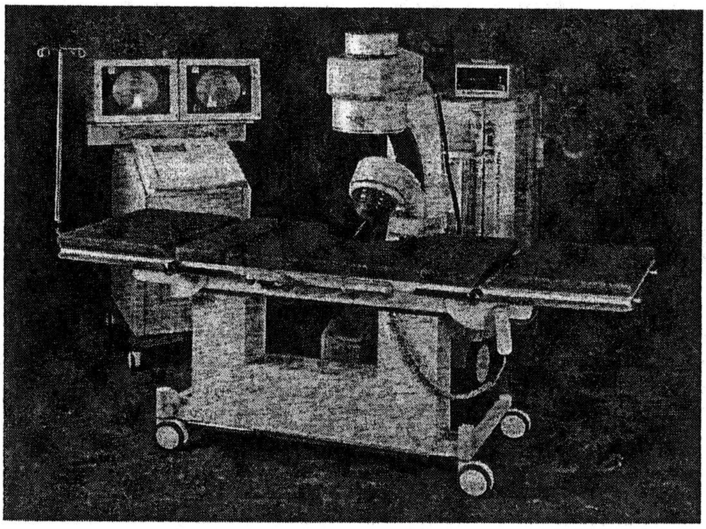

Fig. 5.8 Latest Dornier lithotripter (Compact Delta) for ESWL

ultrasonic lens. It is essential to use underwater SW in the therapy unit because the impedance of human tissue is about the same as that of water. Successful ESWL requires a few hundred shock pulses per session (about 2 hr.), and stone fragments pass the ureter in days or weeks. Note that the peak pressure of the focused SW ranges from 0.2 to 1.15 kb (rise time in 30 to 500 nanoseconds). The gasdynamic focus is a region about 0.1 to 2.5 cm wide, which is near the geometric focal point. At first it was thought that the kidney stone was broken up by shock compression and interface rarefaction, but latest investigations suggest an additional mechanism, viz. hydrodynamic cavitation. When the liquid near the focus of SW is subjected to large tensile stress of rarefaction, the liquid breaks down and vapor-filled cavities form. Under shock compression such cavities collapse, resulting in micro-implosions which pulverize the stone. Today ESWL is practised by urologists worldwide, and multiple thousands of patients are successfully treated each year.

5.7 SW in traffic jams

The congestion of road traffic is a distressful experience of every car driver, but he or she may not be aware of a hidden factor, viz. the SW herein. Before we look into the latter more closely, it is essential to introduce the theory of traffic flow based on the equations below.

$$\frac{\partial \rho}{\partial t} + \frac{\partial q}{\partial x} = 0 \tag{5.21a}$$

$$q = \rho u \tag{5.21b}$$

where ρ is the traffic density (cars/mi), $q = \rho u$ the flow rate (cars/hr), and u the flow velocity (mi/hr). Eq. (5.21b) is an implicit function of density $q = q(\rho)$, viz. the EOS for the theory of traffic flow and SW. The EOS $q = q(\rho)$ is pivotal for the understanding of traffic phenomena as described later. Apparently Eqs. (5.21a) and (5.21b) describe the continuity of traffic flow just like the conservation of mass in gasdynamics:

$$\frac{\partial \rho}{\partial t} + \frac{\partial}{\partial x}(\rho u) = 0$$

On the other hand, Eq. (5.21a) may be written

$$\frac{\partial \rho}{\partial t} + \frac{dq}{d\rho}\frac{\partial \rho}{\partial x} = 0 \qquad (5.21c)$$

which, according to the theory of nonlinear waves, leads to

$$c = \frac{dq}{d\rho} = \text{velocity of density wave} \qquad (5.22a)$$

$$u = q/\rho = \text{car velocity} \qquad (5.22b)$$

$$U = \frac{[q]}{[\rho]} = \frac{\rho_2 u_2 - \rho_1 u_1}{\rho_2 - \rho_1} = \text{shock velocity} \qquad (5.22c)$$

Note that Eqs. (5.21a, b) and (5.22a, b) appear to be analogous to the equations of dispersive waves:

$$\frac{\partial k}{\partial t} + \frac{\partial \omega}{\partial x} = 0$$

$$\omega = \omega(k), \text{ the equation of dispersion}$$

$$c_g = d\omega/dk = \text{group velocity}$$

$$c_f = \omega/k = \text{phase velocity}$$

where k is the wave number ($k = 2\pi/\lambda$, λ being wavelength) and $\omega = 2\pi\nu$, the angular frequency (thus $c_f = \lambda\nu$ in physics). Qualitatively speaking, the density wave on road traffic is dispersive (i.e. the crowding-up and spreading-out of cars).

Noticeably the essence of traffic flow lies in the kinematics of cars' mobility. The nonlinearity of Eq. (5.21c) points to the formation of SW as a result of the steepening density waves, without invoking any Rankine-Hugoniot relations of momentum, energy, and entropy. Gasdynamic SW arise from second-order nonlinear equations, e.g. $c^2 = -v^2(\partial p/\partial v)_s$, $U^2 = v_1{}^2(p_2 - p_1)/v_1 - v_2)$, whereas Eqs. (5.22a) and (5.22c) are correspondingly a first-order nonlinear-wave phenomenon in road traffic. Interestingly Greenshield's model demonstrates this point to a high degree of clarity.

In 1935 Greenshield introduced his empirical EOS

$$q = \rho u = u_m\left(\rho - \rho^2/\rho_j\right) \tag{5.23a}$$

implying the linear relation

$$u = u_m\left(1 - \rho/\rho_j\right) \tag{5.23b}$$

Let us depict these in Fig. 5.9 with the following properties.

(i) $q = u_m\left(\rho - \rho^2/\rho_j\right)$, $q_m = \rho_j u_m/4$, $q_b = q_m/9$

(ii) $q = \rho_j\left(u - u^2/u_m\right)$, $q_m = \rho_j u_m/4$, $q_b = q_m/9$

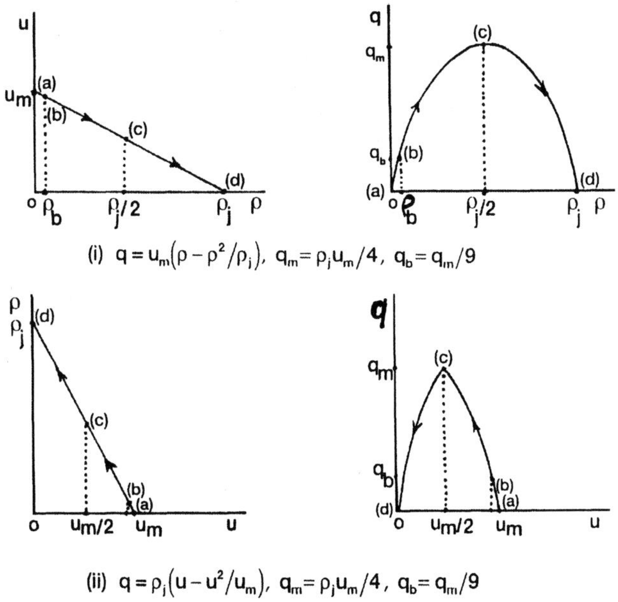

Fig. 5.9 Greenshield model of traffic flow: (i) canonical expressions suitable for SW, (ii) alternative representations featuring flow model only.

State (a) represents the lightest traffic $q = \rho = 0$ at u_m = max. velocity and $(dq/d\rho)_0 = c_0$ = u_f = free speed. State (c) determines the maximum flow $q_m = \rho_j u_m/4$ as a result of $dq/d\rho = 0$. This is the road capacity at best, but it implies some traffic instability with $c = 0$ and $d^2q/d\rho^2 < 0$. State (b) is on the side of light traffic. Let $q_b = q_m/9 = \rho_j u_m/36$. Substituting this in Eq. (5.32a) gives

$$\rho^2 - \rho_j\rho + \rho_j^2/36 = 0$$

$$\rho_b = \left(\frac{1}{2} - \frac{\sqrt{2}}{3}\right)\rho_j$$

$$u_b = \left(\frac{1}{2} + \frac{\sqrt{2}}{3}\right)u_m$$

which determineq_b exactly. State (d) is a traffic jam with $q = u = 0$ at a bumper-to-bumper packing ρ_j. Note also for state (a) $u_f = c_0 = u_m = 4q_m/\rho_j$ = optimal speed for road design. Parameters u_m and ρ_j are essentially empirical constants. In Fig. 5.9 a free-flow or light traffic zone is the segment (a) - (b); segment (b) - (c) marks out the normal traffic zone for flow rate near road capacity; and segment (c) - (d) is the forced-flow or heavy-traffic zone as q and u decrease with increasing ρ. The shock behavior of Greenshield's model is interestingly illustrated below. With $u_m = c_0 = u_f$, as mentioned, Eqs. (5.23b) and (5.22c) may be combined with $x = \rho/\rho_j$ to give

$$U = u_f\left[1 - \rho_j\left(x_2^2 - x_1^2\right)/\left(\rho_2 - \rho_1\right)\right] = u_f\left[1 - \left(x_1 + x_2\right)\right] \tag{5.23c}$$

Thus we have $U = u_f(1 - 2x_1)$ for nearly equal density ratio $x_2 = x_1$, and the SW moves forward with the traffic if the shock path lies on the side (a) - (b) - (c). On the other hand, the SW propagates backward against the traffic if the shock path intersects segment (c) - (d). In a traffic jam, a backward SW is caused by stopping ($u_2 = 0$, $x_2 = 1$, and hence $U = -x_1u_f$). Also SW is produced by starting ($x_1 = 1$ from red to green light): $x_2 = 1 - u_2/u_f$ and $U = -x_2u_f = u_2 - u_f = -u_f/2$ if $u_2 = u_f/2$.

A SW is a sudden jump of traffic density caused by contingency or restriction (accident, patrol demand, road construction, speed limit, no passing, etc.). A sudden release of congestion may cause another SW. A bottleneck is formed when 3 lanes merge into 2 lanes of expressway, causing a backward SW. After this, a reverse shock may ensue to disperse the compressed platoon. In Fig. 5.10 the normal and bottleneck EOS are superimposed for the description of SW caused by trucks which crawl up a 2-lane rural highway (with speed limit of 45 mph) at u_b = 15 mph. A

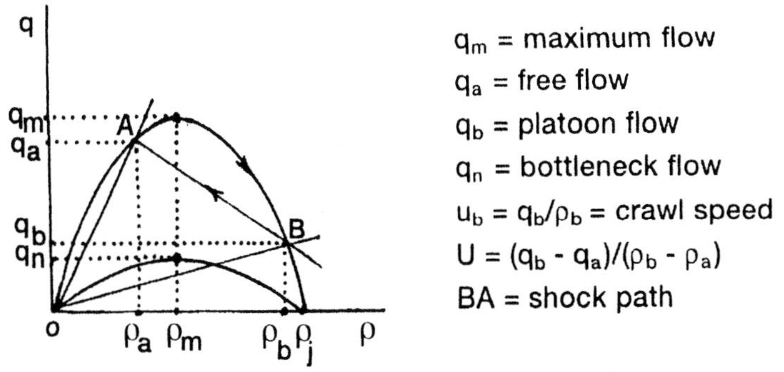

Fig. 5.10 Example of traffic jam and SW.

platoon of cars follow (from A to B) and are forced to slow down. Any additional car may approach this platoon at 45 mph but may suddenly slow down to 15 mph, creating a SW in order to avoid rear-end collision. Now when the truck and platoon travel downhill at 45 mph (not shown in Fig. 5.10), another SW is possible.

There are other EOS $q = q(\rho)$ in use. In 1959 Greenberg introduced an EOS based on NYC Lincoln Tunnel traffic data:

$$q = u_f \rho \ell n \rho_j / \rho$$

which is verifiable by assuming $-\rho du/d\rho$ = constant. Other investigators thought the following EOS to be a better data fit:

$$q = q_m \left(\frac{\rho_j - \rho}{\rho_j - \rho_m} \right) \frac{\rho}{\rho_m}$$

$$q = q_m \left(\frac{\ell n \rho_j / \rho}{\ell n \rho_j / \rho_m} \right) \frac{\rho}{\rho_m}$$

A final remark is worth mentioning. Eq. (5.22c) is based on $\rho_1(U - u_1) = \rho_2(U - u_2)$

with reference to a coordinate frame at rest. This is the ordinary convention versus that of Section 1.4.

5.8 Shock structure and thickness

SW are considered as wave fronts or surfaces of discontinuity in an invicid, non-conducting fluid flow. A geometric surface has area, curvature but no thickness. On the other hand, the conditions of shock jump provide a gross account for the downstream and upstream states only. These two concepts dominate most of the SW theory which has concerned us so far. Perhaps we either overlook or are ignorant of the question: "what happens in the jump or SW per se?" From the viewpoint of molecular kinetics, microscopic processes do occur inside the shock front which is of small thickness. It is the nonequilibrium processes that link up the downstream and upstream states in equilibrium. It is the irreversible processes of entropy production, which convert shock heating into the kinetic energy of molecular random motion. Also the mean free path of molecular collision is a measure of the shock thickness. These are the essence of shock structure and thickness, but further details require laborious manipulations with the Boltzmann equation of gaskinetic theory. In what follows, we seek to begin with a lucid mathematical model.

The Burgers equation

$$\frac{\partial \mu}{\partial t} + \mu \frac{\partial \mu}{\partial x} - \nu \frac{\partial^2 \mu}{\partial x^2} = 0 \tag{5.24a}$$

can serve to describe the velocity profile of a steady shock with velocity U = constant. Let $z = x - Ut$ and $\mu(x, t) = f(z)$. Direct substitution of these into Eq. (5.24a) gives

$$-Uf'(z) + f(z)f'(z) - \nu f''(z) = 0 \tag{5.24b}$$

which is integrable as

$$-Uf + f^2/2 - \nu f' = K = \text{constant}$$

Now it is expedient to re-write the above equation as

$$\frac{df}{dz} = \frac{1}{2\nu}\left(f^2 - 2Uf - 2K\right) = \frac{1}{2\nu}(f - f_1)(f - f_2) \tag{5.24c}$$

with $f_1 = U - \sqrt{U^2 + 2K}$ and $f_2 = U + \sqrt{U^2 + 2K} > f_1$. Integrating again, we obtain

$$\frac{z}{2\nu} = \frac{1}{f_2 - f_1}\,\ell n\left(\frac{f_2 - f}{f - f_1}\right) \quad \text{for } f_1 < f < f_2,\ \mu = \mu_2 \text{ as } z \to -\infty,\ \mu = \mu_1 \text{ as } z \to +\infty$$

which is readily expressed as the solution of Eq. (5.24a):

$$\mu(x,t) = \mu_1 + (\mu_2 - \mu_1)\left\{1 + \exp\left[\frac{\mu_2 - \mu_1}{2\nu}(x - Ut)\right]\right\}^{-1} \quad \text{for } \mu_1 < \mu < \mu_2 \tag{5.24d}$$

Let us plot Eq. (5.24d) in Fig. 5.11 to show the shock structure with thickness

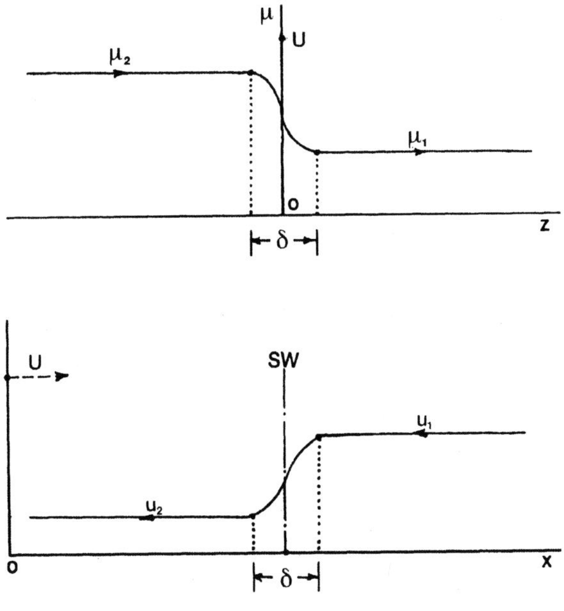

Fig. 5.11 Velocity profile of steady SW.

given by

$$\delta = (\mu_2 - \mu_1)/|d\mu/dz|_{max} = 8\nu/(\mu_2 - \mu_1) \tag{5.24e}$$

Note that $|du/dz|_{max}$ is obtainable from Eqs. (5.24b) and (5.24c) as follows. Letting $f''(z) = 0$ we obtain $f = U$ and hence $\nu f'_{mas} = -K - \frac{U^2}{2} = -\frac{f_1 f_2}{2} - \frac{1}{2}\left(\frac{f_1 + f_2}{2}\right)^2 = \frac{1}{8}(f_1 - f_2)^2$. According to gaskinetic theory, we have the coefficient of viscosity $\eta = 5\pi\rho\ell\hat{c}/32$ with $\hat{c} = (8kT/\pi m)^{1/2}$ = thermal velocity of molecules and ℓ = mean free path. Therefore Eq. (5.24e) with $\nu = \eta/\rho$ predicts $\delta \sim \ell$; and the stronger the shock, the smaller is δ. Apparently Fig. 5.11 is consistent with the convention of shock jumps ($u_2 = U - \mu_2$, $u_1 = U - \mu_1$ and hence $u_2 < u_1 \leq U$; see Figs. 1.1 and 1.2). More justifications can be sought as follows.

In general it is adequate to include viscosity and heat conduction in the basic equations of fluid flow. Thus we may write

$$\rho\left(\frac{\partial u}{\partial t} + u\frac{\partial u}{\partial x}\right) = -\frac{\partial p}{\partial x} + \frac{4}{3}\eta\frac{\partial^2 u}{\partial x^2}$$

for the one-dimensional equation of motion. Apparently Eq. (5.24a) is a simplified version of this by excluding $-\partial p/\partial x$. Moreover, Eq. (5.24a) implies the equation of diffusion if the term $u\partial u/\partial x$ is excluded. In this connection heat conduction should also be included in the energy equation for the study of shock structure. Thus we may generalize Eqs. (1) - (3) in the following forms:

$$\rho u = \rho_o u_o = \text{constant} \tag{5.25a}$$

$$p + \rho u^2 - \frac{4}{3}\eta\frac{du}{dx} = p_o + \rho_o u_o^2 \tag{5.25b}$$

$$\rho u\left(h + \frac{u^2}{2}\right) - \frac{4}{3}\eta u\frac{du}{dx} - \kappa\frac{dT}{dx} = \rho_o u_o\left(h_o + \frac{u_o^2}{2}\right) \tag{5.25c}$$

where the left-hand sides admit variation and gradient within the shock thickness Δx only. The coefficient of thermal conductivity is $\kappa = 15(\gamma - 1)c_v/4$ for $\gamma = 5/3, 7/5$, and $4/3$. It is insightful to consolidate Eqs. (1.8) and (5.25a) - (5.25c) as

$$\rho u T\frac{ds}{dx} = \frac{4}{3}\eta\left(\frac{du}{dx}\right)^2 + \kappa\frac{d^2T}{dx^2} \tag{5.26}$$

by several manipulations. Thus both viscosity and conductivity are the sources of entropy production. Note that Eq. (1.14a) is a rough estimate with $\eta = \kappa = 0$ from a different approach. Through skillful manipulation with Eqs. (5.25a) - (5.25c), other investigators (References 15 and 2) deduce results such as

$$\rho = \rho_1 + (\rho_2 - \rho_1)\left\{1 + \exp\left[c_1\left(M_1^2 - 1\right)(x - Ut)/v'M_1\right]\right\}^{-1} \text{ for } \rho_1 < \rho < \rho_2 \tag{5.27}$$

$$p = \frac{1}{2}(p_2 + p_1) + \frac{1}{2}(p_2 - p_1)\tanh(x/\delta) \text{ for } p_1 < p < p_2 \tag{5.28a}$$

$$\delta = 8av^2/(p_2 - p_1)\left(\partial^2 v/\partial p^2\right)_s = \text{shock thickness } (\delta \sim \ell) \tag{5.28b}$$

where c_1, v', and a are appropriate constants ($M_1 = U/c_1$ = Mach number). Eqs. (5.24c) and (5.27) - (5.28b) suffice to provide a consistent description for the structure and thickness of a steady SW, especially suitable for $M_1 \sim 2$. All such representations are macroscopic (i.e. based on continuum mechanics). Strong SW may have their thickness increased by excitation of freedom-degrees, ionization, radiation, relaxation processes, etc. These effects are more adequately covered in the microscopic theory of gaskinetics. Earlier we have studied ionizing SW, radiating shocks, isothermal shocks, MHD and plasma SW, all with complicated structures. Collisionless plasma shocks are miles thick.

5.9 Computer simulation of shock flow

For an outsider it is boring to play games with numbers, but a new breed of numbers game is very valuable for the insider. This is the numerical modeling or computer simulation of shock flow. In Section 2.5 shock diffraction is highlighted with some results which can only be obtained from shock-tube experiments or numerical modeling (see Fig. 5.13 later). In Section 5.6 it is not feasible to describe the shock focusing for ESWL by simple equations, because the wave patterns are too complex. Thus computer-simulation results are often needed for the design or verification of experiments. Numerical study of ICF seeks to uncover the details and potentiality of laser ablation, shock focusing, and imploding SW. Elsewhere computer experiments for weaponry are especially meritorious (e.g. cost savings, no hardware needed, confidential findings). It is therefore worthwhile to have a glimpse at the modern computational fluid dynamics (CFD) with special emphasis on high-speed flow involving SW.

The following equations suffice to represent the one-dimensional ideal-gas flow (inviscid, non-conducting) in general:

$$\frac{\partial \rho}{\partial t} + \frac{\partial}{\partial x}(\rho u) = 0 \tag{5.29a}$$

$$\frac{\partial}{\partial t}(\rho u) + \frac{\partial}{\partial x}(\rho u u + p) = 0 \tag{5.30a}$$

$$\frac{\partial}{\partial t}\left(\frac{p}{\gamma - 1} + \frac{1}{2}\rho u^2\right) + \frac{\partial}{\partial x}\left(\frac{\gamma p u}{\gamma - 1} + \frac{1}{2}\rho u^3\right) = 0 \tag{5.31a}$$

Note that Eq. (5.31a) is the result of combining the energy equation and EOS:

$$\frac{\partial}{\partial t}\left[\rho\left(E + \frac{1}{2}u^2\right)\right] + \frac{\partial}{\partial x}\left[\rho\left(E + \frac{1}{2}u^2\right)u + pu\right] = 0$$

$$E = p/(\gamma - 1)\rho$$

These equations are not solvable analytically for most problems of shock flow. Then numerical methods and machine calculation are the best resort. Using a finite-difference (FD) grid as shown in Fig. 5.12, we can re-write Eqs. (5.29a) - (5.31a) in FD

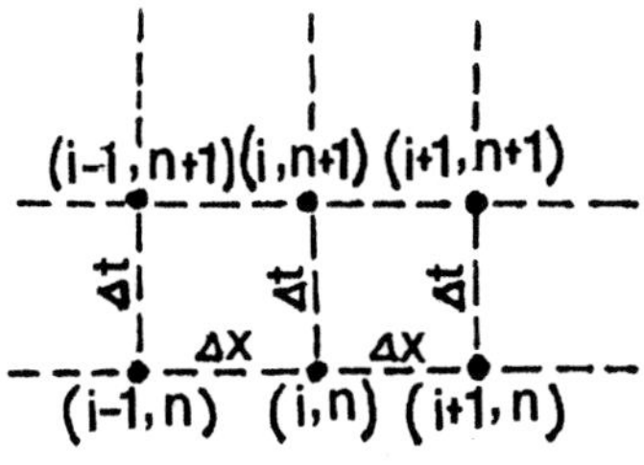

Fig. 5.12 Typical finite-difference grid.

forms below. The algebraic Eqs. (5.29b) - (5.31b) are put in FORTRAN format andstored in the computer. Each grid point (i, n) is located as a mailbox, and data

and

$$\rho_i^{n+1} = \rho_i^n - \left(\frac{\Delta t}{\Delta x}\right)\left[(\rho u)_{i+1/2}^n - (\rho u)_{i-1/2}^n\right] = \mathrm{A}_i^{n+1} \text{ numerically} \tag{5.29b}$$

$$(\rho u)_i^{n+1} = (\rho u)_i^n - \left(\frac{\Delta t}{\Delta x}\right)\left[\left(p+\rho u^2\right)_{i+1/2}^n - \left(p+\rho u^2\right)_{i-1/2}^n\right] = \mathrm{B}_i^{n+1} \tag{5.30b}$$

$$\left(\frac{p}{\gamma-1}+\frac{1}{2}\rho u^2\right)_i^{n+1} = \left(\frac{p}{\gamma-1}+\frac{1}{2}\rho u^2\right)_i^n - \left(\frac{\Delta t}{\Delta x}\right)\left[\left(\frac{\gamma p u}{\gamma-1}+\frac{1}{2}\rho u^3\right)_{i+1/2}^n - \left(\frac{\gamma p u}{\gamma-1}+\frac{1}{2}\rho u^3\right)_{i-1/2}^n\right] = \mathrm{C}_i^{n+1} \tag{5.31b}$$

results (ρ_i^n, p_i^n, u_i^n) are to be stored in or transferred from their pertinent mail boxes for each time step or computational cycle (Δt). Given initial values $\rho_i^0 = \rho(i\Delta x, 0)$, p_i^0, and u_i^0, the computer is instructed to perform calculation of Eqs. (5.29b) - (5.31b) for cycle 1 ($n = 1$, $t_1 = \Delta t$), and solution results $\rho_i^1 = \mathrm{A}_i^1$, $u_i^1 = \mathrm{B}_i^1/\mathrm{A}_i^1$ and $p_i^1 = (\gamma-1)\left(\mathrm{C}_i^1 - \frac{1}{2}\mathrm{A}_i^1\mathrm{B}_i^1\mathrm{B}_i^1\right)$ are stored as the input data for the next cycle ($n = 2$, $t_2 = 2\Delta t$). Since a computer can perform algebraic calculations electronically at high speed, the problem is readily solved in N cycles ($t_0 = 0$, $t_1 = \Delta t$, ... $t_N = N\Delta t$). At the outset the mesh size (Δx) and time step (Δt) are so determined that $(|u|+c)\Delta t/\Delta x < 1$ warrants the accuracy and stability of FD approximation ($c^2 = (dp/d\rho)_s = \gamma p/\rho$). Yet another issue must be addressed now.

When SW form in a supersonic flow, sharp jumps of ρ_i^n, p_i^n, and u_i^n will cause error instability of FD calculations. A remedy is to introduce an artificial viscosity term which will spread the discontinuities over a few meshes. Thus the supersonic flow appears continuous with structured SW just like the effect of real gas discussed in Section 5.8. This viscous pressure is to be given by $q \sim \rho\ell\hat{c}(du/dx)$, which fits the computational need as

$$q = \rho(b\Delta x)^2\left|\frac{du}{dx}\right|\frac{du}{dx} = b^2\rho(\Delta u)^2, \quad b \sim 2$$

or

$$q = b_1\rho c\Delta u, \quad b_1 = 0.05 \text{ to } 0.30$$

In FD form, we have $(\Delta u)^n_{i+1/2} = u^n_{i+1} - u^n_i$ and $(\Delta u)^n_{i-1/2} = u^n_i - u^n_{i-1}$. Now the two expressions for viscous pressures become

$$q^n_{i+1/2} - q^n_{i-1/2} = b^2\left[\rho^n_{i+1/2}\left(u^n_{i+1} - u^n_i\right)^2 - \rho^n_{i-1/2}\left(u^n_i - u^n_{i-1}\right)^2\right] \tag{5.23a}$$

or

$$q^n_{i+1/2} - q^n_{i-1/2} = b_1\left[(\rho c)^n_{i+1/2}\left(u^n_{i+1} - u^n_i\right) - (\rho c)^n_{i-1/2}\left(u^n_i - u^n_{i-1}\right)\right] \tag{5.23b}$$

respectively, and either Eq. (5.32a) or (5.32b) may be included in the bracket of Eq. (5.30b). Likewise, Eq. (5.23c) or (5.32d) should be added correspondingly in the bracket of Eq. (5.31b) in order to improve the accuracy and stability of computation.

The foregoing brief survey outlines the principal methods of CFD in one-dimensional Eulerian formulation. When they are incorporated in a computer with FORTRAN instructions, the whole program is a computer code which can serve to expedite the numerical solution of shock-flow problems. While the fluid particles move from mesh to mesh in a fixed Eulerian FD grid, a Lagrangian flow field is subdivided into a moving FD grid with each mesh always containing the same fluid particle. Thus the Eulerian results appear as cycle-by-cycle snapshots, and the Lagrangian results are visible as a movie story. Each of these has its own advantages and limitations. The FD methods are applicable to two- and three-dimensional flow problems, and more elaborate computer codes are constructed for greater capability. The following list is widely used (circa 1980's) in many research laboratories and graduate schools:

- One-dimensional Lagrangian codes: SIN, WONDY, PUFF, KO, CHART-D, SAP
- Two-dimensional Lagrangian codes: EIC (2DL), TOODY, TROTT, HEMP, STEALTH, PIPE, LEAP, CRAM, NIKE
- Two-dimensional Eulerian codes: 2DE, PIC, OIL, DORF, HELP, HULL, CSQ
- Three-dimensional Eulerian/Lagrangian codes: 3DE, TRIOIL, TRIDORF,

TREEDY, HEMP, STEALTH, METRIC, K3, NIKE, DYNA3D

Today CFD has evolved as an extremely specialized subject, and many more computer codes are available for the solving and simulation of problems. Figs. 5.13 and 5.14 are reproduced from the literature for illustration. The result of shock-tube

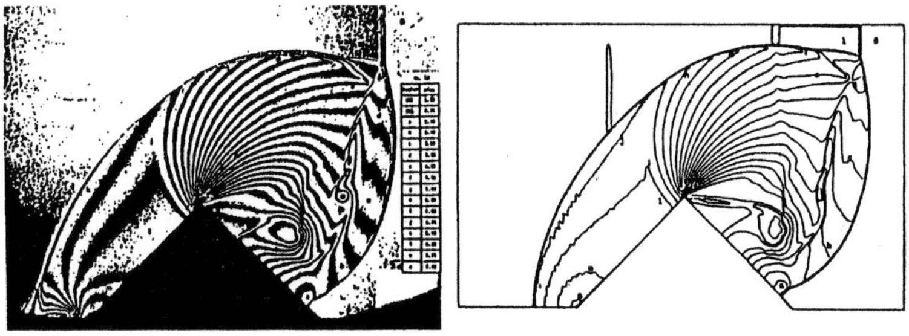

Fig. 5.13 Experimental interferogram vs. computer-simulation results for shock diffraction.

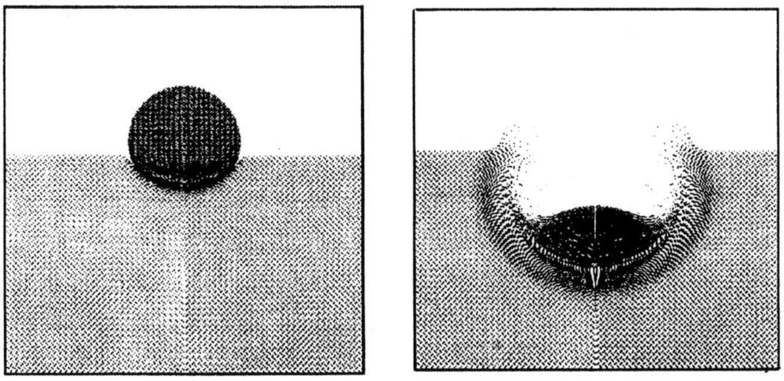

t = 0.25 μsec. t = 1.0 μsec.

Fig. 5.14 Computer-simulation results for meteoric impact.

experiment is compared with that of computer simulation side by side in Fig. 5.13. As tabulated beside the interferometric fringes are the measured density jumps (ratios) from the undisturbed region *o* (ρ_o) to the shocked states 1, *a*, *b*, *c*, *d* (ρ_1, ρ_a, ρ_b, ρ_c, ρ_d), which depict the incident SW and the Mach stem. The reflected shock separates states *f*, *g*, *h*, *i* from state 1, and states *j*, *k*, *l*, *m*, *n* show the diffracted SW.

The slip stream and vortices are noticeable from the swirling pattern and the two eyes. In Fig. 5.14 are shown only two snapshots (t = 0.25 and 1.0 microsecond) which simulate the meteoric-impact cratering at a striking velocity of 30 km/sec. Such a hypervelocity impact generates strong shocks as indicated by the dark contours in both the meteorite and the obstacle, but the computation does not reveal the explosive effects of cratering (i.e. shock melting and vaporization not taken into account by the EOS used).

It is worth noting that elsewhere the method of molecular dynamics (MD) has been developed for the simulation of shock propagation in solids without using the artificial viscosity of CFD. In fact the MD simulation is likewise applicable to fluids. Now considerable effort has to be made to solve the equations of molecular motion

$$md^2r_i/dt^2 = f_i = -d\varphi/dr_i, \quad i = 1, 2, 3 \ldots N \tag{5.32}$$

in finite-difference scheme with N = 100 to 1000 molecules, but super computers can deal with N = 10^6 molecules for given initial and boundary conditions. In crystalline solids each lattice point of mass m will undergo one-, two- or three-dimensional nonlinear vibrations due to the interaction potential φ (say, Lennard-Jones or Morse potential function). The numerical solution $r_i = r_i(t)$ is then the data source for computing velocities, forces, energies, material properties, and EOS which are needed to determine the shock propagation or shock Hugoniot for solids. Such procedures are also referred to as the lattice dynamics method. Apparently MD is a microscopic/discrete model, and CFD is a macroscopic/continuum model. In contrast to the finite-difference method of CFD, a finite-element method (FEM) is very popular for the numerical analysis of structural vibrations/dynamics. Here we do not intend to include the ramification of FEM in CFD. Yet another model is the Monte Carlo simulation sophisticatedly based on the probabilistic theory of random walk.

Epilogue

This monograph is an attempt to update the many aspects of SW and EOS, which are of practical value in science and technology. A style of brevity and adaptability has been deliberated comprehensively. The five chapters of this short handbook should prove to be all essential for an expedient grasp of the behavior and properties of SW in various forms (planar, normal, oblique, cylindrical, spherical, strong, and weak). Also covered are the great majority of EOS for SW study, a theory of SW in seven particular environments, and nine special topics which may fill a good many volumes. Note that analytically simple and useful are the EOS of ideal gas (quantum including classical as an extreme limit), van der Waals, and Grüneisen (especially with the linear shock Hugoniot). The modified EOS of classical ideal gas are the key to the understanding of the seven interdisciplinary and nine special topics. Yet there are still many SW and EOS beyond the scope of this book; they are oblique MHD shocks, collisionless SW in the outer space and from the sun, relativistic SW, nuclear EOS, etc. and advanced studies are the key to these specialties.

The science and technology of SW may be explored from three main approaches: (a) the theory of nonlinear waves; (b) the shock-tube, light-gas-gun, wave-lens (contact explosion), and other experiments; and (c) numerical simulation by CFD or MD. This book lacks much of these textbook-like approaches. The nonlinear-wave theory is expounded in the texts of applied mathematics and mechanics which deal with hyperbolic partial differential equations, compressible fluid flows, and finite deformation-rates of solids at high pressure. It would be too inconclusive to take up this academic topic in our monograph. The other two approaches are apparently beyond the scope of this monograph. So our approach is meant to be pragmatically analytic; we bring SW and EOS cohesively together; and the coverage of methodology and topics is modest for a short handbook. This is the nature of our monograph, which differs from typical textbooks.

The backbone of SW analysis is the Rankine-Hugoniot jump conditions and EOS of consequential responses. Here the wave carrier, momentum, energy, and properties of matter constitute the generation and support of SW with the effects of supersonic speeds, high temperature, and/or high pressure. The four states of matter provide the environments which carry or contain the SW. Yet another environment is prone to SW without the impetus of momentum and energy. It is the contingency or choked flow on road traffic, that causes SW to propagate backward or forward. These are real SW with their formation and propagation attributable to EOS or models of traffic flow, but they are kinematic SW which lack any link to the momentum and energy relations of Rankine and Hugoniot.

It appears that shock phenomena occur in Rankine-Hugoniot or non-Rankine-Hugoniot environments. Interestingly enough, SW and EOS are intrinsically complementary to each other. Our inquiry need not end here. Hopefully, new kinds of shock wave and environment might emerge!

References

(1) A. Sommerfeld: Mechanics of deformable bodies (Academic Press, 1964), 396pp.

(2) L.D. Landau and E.M. Lifshitz: Fluid mechanics (Pergamon Press, 1959), 536pp.

(3) F.H. Harlow and A.A. Amsden: Fluid dynamics, a LASL monograph LA-4700 (Los Alamos Scientific Laboratory, 1971), 115pp.

(4) A. Sommerfeld: Thermodynamics and statistical mechanics (Academic Press, 1964), 401pp.

(5) M.W. Zemansky: Heat and thermodynamics (McGraw Hill, 1968), 658pp.

(6) R. Becker: Theory of heat (Springer-Verlag, 1967), 380pp.

(7) L.D. Landau and E.M. Lifshitz: Statistical physics (Pergamon Press, 1958), 484pp.

(8) S. Eliezer, A. Ghatak, and H. Hora: An introduction of equations of state, theory and applications (Cambridge University Press, 1986), 366pp.

(9) V.N. Zharkov and V.A. Kalinin: Equations of state for solids at high pressures and temperatures (Consultants Bureau, 1971), 257pp.

(10) J.V. Sengers, R.E. Kayser, C.J. Peters, and H.J. White, Jr. (ed.): "Equations of state for fluids and fluid mixtures (Elsevier, 2000), Part I and II, 885pp.

(11) S.P. Marsh (ed.): LASL shock Hugoniot data (University of California Press, 1980), 658pp.

(12) R.F. Trunin: Shock compression of condensed materials (Cambridge University Press, 1998), 167pp.

(13) H.W. Liepmann and A. Roshko: Elements of gasdynamics (Wiley, 1957), 439pp.

(14) R. Courant and K.O. Friedrichs: Supersonic flow and shock waves (Interscience Publishers, 1948), 464pp.

(15) Ya. B. Zeldovich and Yu. P. Raizer: Physics of shock waves and high temperature hydrodynamic phenomena (Academic Press, 1967), vol I and II, 916pp.

(16) J.N. Bradley: Shock waves in chemistry and physics (Methuen, 1962), 370pp.

(17) J.K. Wright: Shock tubes (Methuen, 1961), 164pp.

(18) I.I. Glass and J.P. Sislian: Nonstationary flows and shock waves (Oxford University Press, 1994), 561pp.

(19) G. Ben-Dor: Shock wave reflection phenomena (Springer-Verlag, 1992), 307pp.

(20) G.B. Whitham: Linear and nonlinear waves (Wiley, 1974), 636pp.

(21) K.P. Stanyukovich: Unsteady motion of continuous media (Pergamon Press, 1960)

745pp.

(22) V.C.A. Ferraro and C. Plumpton: An introduction to magnetofluid mechanics (Oxford University Press, 1966), 254pp.

(23) R. Kinslow (ed.): High-velocity impact phenomena (Academic Press, 1970), 579pp.

(24) P. Caldiro and H. Knoepfel (ed.): Physics of high energy density (Academic Press, 1971), 418pp.

(25) R.H. Cole: Underwater explosions (Dover paperback, 1965), 427pp.

(26) S. Glasstone and P.J. Dolan: The effects of nuclear weapons (U.S. Government Printing Office, 1977), 653pp.

(27) G. Velarde, Y. Ronen, and J.M. Martinez-Val (ed.): Nuclear fusion by inertial confinement (CRC Press, 1993), 751pp.

(28) J.D. Lindl: Inertial confinement fusion (Springer-Verlag, 1998), 204pp.

(29) W. Grüner and H. Stöcker (ed.): The nuclear equation of state, Part A, discovery of nuclear shock waves and the equation of state (Plenum Press, 1989), 802pp.

(30) W.A. Ashton: The theory of road traffic flow (Methuen, 1966), 178pp.

The following symposia are worth noting

(1) Proceedings of the first shock tube symposium (1957)
(2) Proceedings of the second shock tube symposium (1958)
(3) Proceedings of the third shock tube symposium (1959)
(4) Proceedings of the fourth shock tube symposium (1961)
(5) Proceedings of the fifth international shock tube symposium (1966)
(6) Proceedings of the sixth international shock tube symposium (1969)
(7) Proceedings of the seventh international shock tube symposium (1970)
(8) Proceedings of the eighth international shock tube symposium (1971)
(9) Proceedings of the ninth international shock tube symposium (1973)
(10) Proceedings of the tenth international shock tube symposium (1975)
(11) Proceedings of the eleventh international symposium on shock tubes and waves (1978)
(12) Proceedings of the twelfth international symposium on shock tubes and waves (1980)
(13) Proceedings of the thirteenth international symposium on shock tubes and waves (1982)
(14) Proceedings of the fourteenth international symposium on shock tubes and waves (1984)
(15) Proceedings of the fifteenth international symposium on shock waves and shock tubes (1986)
(16) Proceedings of the sixteenth international symposium on shock waves and shock tubes (1987)
(17) Proceedings of the seventeenth international symposium on shock waves and shock tubes (1990)
(18) Proceedings of the eighteenth international symposium on shock waves (1991)
(19) Proceedings of the nineteenth international symposium on shock waves (1993)
(20) Proceedings of the twentieth international symposium on shock waves (1995)
(21) Proceedings of the twenty-first international symposium on shock waves (1997)
(22) Proceedings of the twenty-second international symposium on shock waves (1999)
(23) Proceedings of the twenty-third international symposium on shock waves (2001)

The following proceedings are also worth noting

(1) Shock waves in condensed matter I (1979)
(2) Shock waves in condensed matter II (1981)
(3) Shock waves in condensed matter III (1983)
(4) Shock waves in condensed matter IV (1985)
(5) Shock waves in condensed matter V (1987)
(6) Shock waves in condensed matter VI (1989)
(7) Shock compression of condensed matter VII (1991)
(8) Shock compression of condensed matter VIII (1993)
(9) Shock compression of condensed matter IX (1995)
(10) Shock compression of condensed matter XI (1997)
(11) Shock compression of condensed matter XI (1999)
(12) Shock compression of condensed matter XII (2001)

Appendix

The following are reprints of the author's earlier publications to supplement Chapter 3.

1. "Shock-wave behavior and properties of solids," in *J. Appl. Phys.* 42, 3212 (1971)
2. "Note on shock compression of solids," in *J. Appl. Phys.* 45, 2346 (1974)
3. "A third-order adiabat for solids at high pressure," in *J. Appl. Phys.* 46, 3672 (1975)
4. "The generalized compressibility equation of Tait for dense matter," in *J. Phys. D: Appl. Phys.* 7, 158 (1974)
5. "A special class of ideal quantum gases," in *Am. J. Phys.* 40, 1261 (1972)
6. "Analytical model for super-ideal gases,: in *Foundations Phys.* 4, 207 (1974)
7. "Thermodynamics of shock compression of metals," in *J. Chem. Phys.* 45, 1979 (1966)
8. "Shock dynamics of hypervelocity impact of metals," in *J. Franklin Inst.* 276, 39 (1963)

Acknowledgment

Many thanks are due to American Institute of Physics for permissions to reproduce the above items No. 1, 2, 3, 5, and 7 from the respective journals. Special permission is granted by Institute of Physics Publishing Ltd., UK for the reproduction of No. 4 (with the author's indebtedness, the journal's homepage at Internet www.iop.org/journals/jphysd). Also the author is grateful to Plenum Publishing Corp and the Franklin Institute for their permissions to reproduce No. 6 and No. 8 respectively.

Reprinted from:
JOURNAL OF APPLIED PHYSICS VOLUME 42, NUMBER 8 JULY 1971

Shock-Wave Behavior and Properties of Solids

Y.K. Huang
Research Laboratory, Watervliet Arsenal, Watervliet, New York 12189
(Received 10 December 1970)

Shock-wave behavior and properties of solids are described on the basis of a quadratic representation for the shock velocity and with results of considerable generality and simplicity. An implicit treatment is provided by coupling the Hugoniot adiabat with the isentrope in the Grüneisen equation of state for solids. This also leads to correlations between data and results of shock and hydrostatic measurements for high pressures.

I. INTRODUCTION

During the past two decades or so, shock waves in solids as a research tool[1,2] have greatly enhanced our understanding of high-pressure physics. It is of profound interest to consider the general behavior and properties of shock waves. From the experimental point of view, there exists a linear relationship between the shock and material velocities in solids. This and the three equations for the conservation of mass, momentum, and energy suffice to provide a phenomenological description for the shock compression of solids. Further correlations can be established with the aid of thermodynamics. As a result, calculations and interpretations have been made to fit to pressures ranging from a few hundred kbar to 2 Mbar.[3] (Higher pressures have been reported by Soviet research workers.)[2,4] From the analytical point of view, several investigators[5,6] have proposed some theoretical derivations for the linear law of shock velocity. It may be noted that the essence of such exposition lies in either the Grüneisen law or the nonlinear compressibility of solids. This point we have demonstrated clearly in a recent paper[7] concerning some high-pressure properties of solids.

For still higher pressures, say above 10 Mbar, a quadratic equation has been considered to be the better representation of strong shock-wave behavior for solids. In this connection, there have also appeared two papers[8,9] which offer some analytical interpretation for the parabolic law of shock velocity. It seems to the author that, with the pertinent physical validity kept in mind, considerable simplicity and insight can be gained from this basic investigation. Instead of deriving the parabolic law, we seek to make use of it in this paper for a generalized analysis of the shock-wave behavior and properties of solids. Our formulation is based on the coupling of the Hugoniot adiabat with the isentrope in the Grüneisen equation of state for solids. Through a few basic relations of thermodynamics, an interplay is exhibited between shock and hydrostatic compressions. Such an informative link is a particular interest to modern research of high pressures. In view of the pertinent physical validity, we have made our analysis tractable to the fourth-order pressure derivatives only, and our results have consistently generalized those of Refs. 1 and 3.

II. SHOCK-DYNAMIC DESCRIPTION

The shock-wave behavior of solids has been described, more or less completely, in the literature by several leading investigators[1-4] of this field. It is still of interest to consider some fundamental properties of solids as deducible from shock-wave measurements. For this purpose, we shall summarize and generalize the standard representation[3,9] of shock-wave behavior as follows:

$$U = a_0 + a_1 u - a_2 u^2, \tag{1}$$

$$\epsilon = u/U = 1 - v/v_0, \tag{2}$$

$$p_H = uU/v_0 = \epsilon U^2/v_0, \tag{3}$$

where U denotes the shock velocity, u is the material velocity, ϵ is the relative compression, v is the specific volume (v_0 is its initial value at zero pressure and room temperature), p_H is the shock pressure, and a_0, a_1, and a_2 are the three experimental constants. These equations are based on both experiment and theory, and they can be combined into a single equation known as the shock adiabat. For the sake of simplicity, as well as physical validity, we shall always deal with real numbers only. Thus, we may combine Eqs. (1) and (2) to yield

$$(a_2\epsilon^2)^2 U^4 - [(1 - a_1\epsilon)^2 + 2a_0 a_2 \epsilon^2] U^2 + a_0^2 = 0. \tag{4}$$

It should be noted that the first term of Eq. (4) is $(a_2u^2)^2$. Because of the experimental nature of Eq. (1), a_2 is a small positive number to exhibit a slight curvature for the representation with higher precision. Therefore, the small quantity a_2u^2 will concern us more seriously than the higher-order term $(a_2u^2)^2$. Dropping out the first term of Eq. (4), we obtain

$$U^2 = a_0^2[(1 - a_1\epsilon)^2 + 2a_0a_2\epsilon^2]^{-1}. \tag{5}$$

Substituting Eqs. (2) and (5) into Eq. (3), we get the shock adiabat

$$p_H(v) = a_0^2(v_0 - v)\{[v_0 - a_1(v_0 - v)]^2 + 2a_0a_2(v_0 - v)^2\}^{-1}, \tag{6}$$

which has been of chief concern to Prieto and Renero[9] in the mathematical jargon of complex numbers.

It will become clear that Eq. (6) plays an important role in the investigation of shock waves and high pressures. As the Los Alamos scientists[1,3] have used the third-order pressure derivatives with a two-parameter shock adiabat, so we seek to consider here a three-parameter description tractable up to the fourth order. Thus, the fundamental properties of the shock adiabat may be described with

$$p_H'(v_0) = -a_0^2/v_0^2, \tag{6a}$$

$$p_H''(v_0) = 4a_0^2 a_1/v_0^3, \tag{6b}$$

$$p_H'''(v_0) = -6a_0^2(3a_1^2 - 2a_0 a_2)/v_0^4, \tag{6c}$$

$$p_H''''(v_0) = 96a_0^2(a_1^3 - 2a_0 a_1 a_2)/v_0^5, \tag{6d}$$

where the single and multiple primes denote derivatives of pertinent orders. Through a few basic correlations to be introduced later, these shock-dynamic properties will be used as our main sources of input data.

III. QUASISTATIC REPRESENTATION

Strictly speaking, shock compression of solid is an irreversible process whose end states are correlated by the Hugoniot equation

$$E - E_0 = \tfrac{1}{2}(v_0 - v)p_H, \tag{7}$$

where E denotes the specific internal energy and subscript 0 refers to the initial state of zero pressure and room temperature. This finite-difference equation of Hugoniot is comparable with the differential form for a reversible adiabatic compression. It may be pointed out that the period of shock transition is a few orders of magnitude longer than the relaxation time of most nonequilibrium processes. For these reasons we shall adopt a quasi-equilibrium approach for the shocked state of the solid. In other words, we can determine a complete thermodynamic representation[10] for it as $p_H = p_H(v)$, $E = E(v)$, $S = S(v)$, and $T = T(v)$, S and T being the entropy and temperature, respectively. Such a description is amenable to numerical purposes.

For the interest of analytical simplicity, it is far reaching to use the Grüneisen equation of state for solids[1,11]

$$pv = \gamma E + \varphi(v), \tag{8}$$

where γ is the Grüneisen parameter and $\varphi(v)$ is some arbitrary function to account for the interatomic forces in the solid. However, we shall not consider $\varphi(v)$ as it should be treated in lattice dynamics and solid-state physics. Without loss of generality but with considerable gain in simplicity and clarity, we shall eliminate $\varphi(v)$ from Eq. (8) and form the coupling as follows[1,7]

$$p_H - p_s = (E - E_s)\gamma/v \tag{9}$$

with

$$p_s = -dE_s/dv \tag{10}$$

denoting the isentrope corresponding to the shock adiabat with the same volume. Eliminating E between Eqs. (7) and (9) yields

$$[1 + \tfrac{1}{2}(v - v_0)\gamma/v]p_H = p_s - (E_s - E_0)\gamma/v. \tag{11}$$

This coupled equation of state is useful on several occasions, but we need only to consider here some of its important properties. Although it is rather laborious, successive differentiation of Eq. (11) with appropriate substitution yields

$$p_s'(v_0) = p_H'(v_0), \tag{11a}$$

$$p_s''(v_0) = p_H''(v_0), \tag{11b}$$

$$p_s'''(v_0) = p_H'''(v_0) + \tfrac{1}{2}p_H''(v_0)\gamma_0/v_0, \tag{11c}$$

$$p_s''''(v_0) = p_H''''(v_0) + p_H'''(v_0)\gamma_0/v_0 + 2[v_0\gamma_0' - \gamma_0(1 + \tfrac{1}{4}\gamma_0)]p_H''(v_0)/v_0^2, \tag{11d}$$

These interrelations will be used to link shock and hydrostatic compressions. Before proceeding further, we shall have to dwell on a few problems of thermodynamics.

IV. THERMODYNAMIC CONSIDERATION

From thermodynamics[11,12] we have the basic relation

$$p_s'(v)/p_i'(v) = c_p/c_v = 1 + \gamma\alpha T, \tag{12}$$

where $p_i = p_i(v)$ denotes the isotherm, c_p and c_v are the two specific heats, and α is the coefficient of thermal expansion. Under normal conditions solids behave more or less with $c_p \approx c_v$ and $(\gamma\alpha T)_0 \ll 1$. Also, the degenerate state of solids may be looked upon as a superideal gas with $(\alpha T)_\infty \to 1$ and $\gamma \to \gamma_\infty$ as p_s, $p_i \to \infty$. As a result of these, we deduce through logarithmic differentiation

$$p_s''(v)/p_s'(v) = p_i''(v)/p_i'(v), \tag{12a}$$

$$p_s'''(v)/p_s'(v) = p_i'''(v)/p_i'(v), \tag{12b}$$

$$p_s''''(v)/p_s'(v) = p_i''''(v)/p_i'(v). \tag{12c}$$

Obviously, we may write $p_s(v_0) = p_i(v_0) = 0$, $p_s'(v_0) = p_i'(v_0)$, etc. In other words, the isentrope and the isotherm are in fourth-order contact, whereas the Hugoniot and the isentrope have a second-order tangency at the initial point [see Eqs. (11a) and (11b)]. Such a second-order contact is the necessary and sufficient conditions for our quasistatic representation or shock stability.[13] Also, the interrelationship among p_H, p_s, and p_i will be especially useful later.

It will be convenient to consider K, the reciprocal of isothermal compressibility, as an arbitrary function of pressure:

$$K = K(p_i). \tag{13}$$

For solids, this is an intricate nonlinear function However, we can explore the general behavior and properties of the arbitrary function from the following relations:

$$p_t' = -K/v, \qquad (13a)$$

$$p_t'' = K(K'+1)/v^2, \qquad (13b)$$

$$p_t''' = -K[KK'' + (K'+3)K' + 2]/v^3, \qquad (13c)$$

$$p_t'''' = K[K^2K''' + 2K(2K'+3)K'' + (K')^3 + 6(K')^2 + 11K' + 6]/v^4. \qquad (13d)$$

These have an important bearing upon the Grüneisen law $\gamma = \gamma(v)$.

The parameter γ of Eqs. (8) and (11) may be evaluated by the use of either the Slater formula[14]

$$\gamma_S = -\tfrac{1}{2}vp_t''/p_t' - \tfrac{2}{3} = \tfrac{1}{2}K' - \tfrac{1}{6} \qquad (14)$$

or the Dugdale-MacDonald formula[1,3,15]

$$\gamma_{D-M} = -\tfrac{1}{2}v(p_t v^{2/3})''/(p_t v^{2/3})' - \tfrac{1}{3} = \tfrac{1}{2}K(3K'-2)/(3K-2p_t) - \tfrac{1}{6} \qquad (15)$$

Although they yield initial values with a difference of $\frac{1}{3}$, both Eqs. (14) and (15) lead to the same limiting value as p_t, $K \to \infty$. As pointed out earlier[7] there arose considerable controversy over that initial difference, but Rodionov[16] has recently shown an almost 50-50 chance for the fitness of Eqs. (14) and (15). Thus we are led to the opinion that both the Dugdale-MacDonald and the Slater formulas have merits of their own and that their constitutive differences should be attributed to the different types of solids which they can suit best. In other words, both formulas are equally useful for the right kinds of solids. With the simpler formula of Slater, we deduce[7] the salient characteristics of $\gamma = \gamma(v)$ and $K = K(p_t)$ as follows:

$$\gamma > 0, \quad \gamma' > 0, \quad \gamma'' > 0 \qquad (16)$$

$$K > 0, \quad K' > 0, \quad K'' < 0, \quad K''' > 0. \qquad (17)$$

These inequalities are composed of pressure derivatives up to the fourth order, and their validity implies the underlying criteria of thermodynamic and shock stabilities.

V. HIGH-PRESSURE PROPERTIES

At this point we may use either Eq. (14) or (15) in combination with the appropriate equations of Secs. II, III, and IV to determine some fundamental properties of solids under static or dynamic compression. Let us first proceed with the Slater formula and make substitution at the initial state. Thus, from Eqs. (14), (12a), (11a), (11b), (6a), and (6b), we get

$$\gamma_0 = 2a_1 - \tfrac{2}{3}. \qquad (14a)$$

Taking similar steps, we obtain these values in sequence:

$$p_s'''(v_0) = -2a_0^2(7a_1^2 + 2a_1/3 - 6a_0a_2)/v_0^4, \qquad (11c')$$

$$\gamma_0' = (a_1^2 + 4a_1/3 + 6a_0a_2)/v_0, \qquad (14b)$$

$$p_s''''(v_0) = 4a_0^2(15a_1^3 + 3a_1^2 + 10a_1/9 - 2a_0a_2 - 30a_0a_1a_2)/v_0^5, \qquad (11d')$$

$$\gamma_0'' = 2(5a_1^3 + 4a_1/9 + 4a_0a_2 + 6a_0a_1a_2)/v_0^2, \qquad (14c)$$

$$K_0 = a_0^2/v_0, \qquad (13a')$$

$$K_0' = 4a_1 - 1, \qquad (13b')$$

$$K_0'' = -2v_0(a_1^2 + 4a_1/3 + 6a_0a_2)/a_0^2, \qquad (13c')$$

$$K_0''' = 4v_0^2(7a_1^3 + 8a_1^2/3 + 4a_1/9 + 4a_0a_2 + 18a_0a_1a_2)/v_0^4 \qquad (13d')$$

Rather tedious manipulation is involved with the Dugdale-MacDonald formula and associated equations. So we give only the pertinent results as follows:

$$\gamma_0 = 2a_1 - 1, \qquad (15a)$$

$$p_s'''(v_0) = -2a_0^2(7a_1^2 + a_1 - 6a_0a_2)/v_0^4, \qquad (11c'')$$

$$\gamma_0' = (a_1^2 - a_1/3 + 6a_0a_2 + 5/9)/v_0, \qquad (15b)$$

$$p_s''''(v_0) = 2a_0^2(30a_1^3 + 11a_1^2/3 + 47a_1/9 - 6a_0a_2 - 60a_0a_1a_2)/v_0^5, \qquad (11d'')$$

$$\gamma_0'' = (10a_1^3 - 13a_1^2 + 77a_1/9 - 2a_0a_2 + 12a_0a_1a_2 - \tfrac{50}{27})/v_0^2, \qquad (15c)$$

$$K_0 = a_0^2/v_0, \qquad (13a'')$$

$$K_0' = 4a_1 - 1, \qquad (13b'')$$

$$K_0'' = -2v_0(a_1^2 + a_1 + 6a_0a_2)/a_0^2, \qquad (13c'')$$

$$K_0''' = 2v_0^2(14a_1^3 - 7a_1^2/3 + 29a_1/9 + 6a_0a_2 + 36a_0a_1a_2)/a_0^4. \qquad (13d'')$$

From the above two sets of results we observe (a) that both inequalities (16) and (17) are verified; (b) the close similarity between the Slater and Dugdale-MacDonald deductions; and (c) that some isentropic and isothermal compression properties are directly related to shock-compression data. With $a_2 = 0$, in particular, our results reduce to those as implied and partly given in Refs. 1–3. It may be noted that Eqs. (15a) and (15b) check with the results of Los Alamos.[1,3] Also, our Eqs. (13c′) and (13c″) check well with Ruoff's Eq. (26) in Ref. 6, provided that the appropriate γ_0 is substituted in his Eq. (26) disregarding the difference between isentropic and isothermal bulk moduli at the initial state. Here we observe some results from two different approaches to be in excellent agreement.

VI. CONCLUDING REMARK

This investigation has generalized some methods and results as given in Refs. 1–3 and 6 for shock waves in solids. From the macroscopic point of view, the initial state we choose is simpler and more natural than that associated with the 0 °K isotherm.[1,8] It is our primary concern to explore the fundamental properties of solids under static or dynamic compression. Within the framework of our representation, both shock-dynamic and hydrostatic compression properties are interrelated with pressure derivatives up to the fourth order. Our explicit results should be of complementary interest to investigators who are conversant with the literature concerning shock compression of solids.

[1]M. H. Rice, R. G. McQueen, and J. M. Walsh, Solid State Phys. 6, 1 (1958).
[2]L. V. Al'tshuler, Sov. Phys. Usp. 8, 52 (1965).
[3]R. G. McQueen and S. P. Marsh, J. Appl. Phys. 31, 1253 (1960).
[4]L. V. Al'tshuler, B. N. Moiseev, L. V. Popov, G. U. Simakov, and R. F. Trunin, Sov. Phys. JETP 27, 420 (1968).
[5]J. Berger and S. Joigneau, Compt. Rend. 249, 2506 (1959).
[6]A. L. Ruoff, J. Appl. Phys. 38, 4976 (1967).
[7]Y. K. Huang, J. Chem. Phys. 53, 571 (1970).
[8]D. J. Pastine and D. Piacesi, J. Phys. Chem. Solids 27, 1783 (1966).
[9]F. E. Prieto and C. Renero, J. Appl. Phys. 41, 3876 (1970).
[10]Y. K. Huang, J. Chem. Phys. 46, 4570 (1967); Y. K. Huang, in *Colloque International du C. N. R. S. sur les Proprietes Physiques des Solides sous Pression* (Grenoble, France, 1969), pp. 43—47.
[11]E. Grüneisen, *The State of A Solid Body*, NASA Republication RE-2-18-59W (U. S. GPO, Washington, D. C., 1959).
[12]M. W. Zemansky, *Heat and Thermodynamics* (McGraw-Hill, New York, 1957), 4th ed., pp. 251, 253.
[13]K. P. Stanyukovich, *Unsteady Motion of Continuous Media* (Pergamon, London, 1960), pp. 203, 209.
[14]J. C. Slater, *Introduction to Chemical Physics* (McGraw-Hill, New York, 1939), p. 239.
[15]J. S. Dugdale and D. K. C. MacDonald, Phys. Rev. 89, 832 (1953).
[16]K. P. Rodionov, Phys. Metals Metallog. 23, 44 (1967).

Note on shock compression of solids

Y. K. Huang

Watervliet Arsenal, Watervliet, New York 12189
(Received 18 December 1973)

In the framework of a simple compression wave and a weak shock wave, an isentrope of the Tait type and a Hugoniot adiabat based on the linear velocity relation are shown to verify each other consistently. Thus, the two empirical equations of state are given a semianalytical treatment on favorable terms.

Shock compression of solids can be achieved either by contact explosion or by hypervelocity impact, with peak pressures ranging from a few hundred kilobars to tens of megabars.[1,2] Such high pressures are quasi-statically sustained for a couple of microseconds or multinanoseconds only. Thereafter, shock attenuation will be more intricate to analyze. It should be noted that the quasiequilibrium approach has proved adequate for determining some fundamental properties of shock waves and the equations of state of solids. In this paper we will give a semianalytical consideration for the basic representation of results from shock-wave measurements of solids.

Let us refer to the linear relation[1]

$$U=a_0+a_1u, \tag{1}$$

where U is the shock-wave velocity, u is the particle velocity, and a_0 and a_1 are two empirical (positive) constants whose physical meaning will become apparent later in the discussion to follow. From Eq. (1) and the conservation of momentum we can write the Hugoniot adiabat as

$$p_H(u)=\rho_0uU=\rho_0a_0u+a_1\rho_0u^2, \tag{2}$$

with p_H denoting the shock pressure and ρ_0 denoting the initial density of the solid. With Eq. (1), $v=1/\rho$, and $U/v_0=(U-u)/v$, Eq. (2) can readily be transformed into

$$p_H(v)=a_0^2(v_0-v)[v_0-a_1(v_0-v)]^{-2} \tag{3}$$

which has the following initial values: $p_H(v_0)=0$, $p_H'(v_0)=-(a_0/v_0)^2$, and $p_H''(v_0)=4a_1a_0^2/v_0^3$. At this point a few comments are of relevance. In the theory of detonation[3] we have the Chapman-Jouguet condition given by $U=u+a$. Since a detonation wave consists of a shock front and an energy-releasing zone in dense matter, it seems plausible to consider the local sound velocity as $a=a_0+a_0'u$ which leads to Eq. (1) with $a_1=1+a_0'$. On the other hand, Eq. (2) appears to be an interpolation formula between two extreme cases. For very weak shocks with $u\ll a_0$ or $a_1=0$, Eq. (2) reduces to the momentum equation of linear waves with characteristic impedance $Z_0=\rho_0a_0$. For very strong shocks with $u\gg a_0$, Eq. (2) becomes one for the ideally shaped charge (metal) jet with $a_1=\frac{1}{2}$. From Eq. (3) we can arrive at the limit of shock infinity with $v_0/v_\infty=a_1/(a_1-1)$, as in the case of a polytropic gas. Thus, Eq. (3) is not only monotonic in $v_0\geq v\geq v_\infty$ but also asymptotically polytropic.

In the literature there are at least four papers[4-7] which have provided some analytical interpretations for Eq. (1). These interpretations are based on different models or approximations, and they do not constitute a unified treatment. For this reason and others, it is worthwhile for us to offer a new analysis as follows. In view of the proximity of weak shock wave to simple compression wave, we may write

$$\frac{-dp}{dv}=\left(\frac{dp}{du}\right)^2 \tag{4}$$

for both reversible and quasireversible (negligible irreversibility) adiabats.[3,8] Physically, Eq. (4) may be looked upon as the equation of acoustic impedance $Z=\rho a=(-dp/dv)^{1/2}$. With $Z=Z(p)$, Eq. (4) can serve to transform $p=p(v)$ into $p=p(u)$ and vice versa. It is clear that Eq. (4) implies $dp/dv<0$, $d^2p/dv^2=2(dp/du)^2(d^2p/du^2)$, and hence $d^2p/du^2>0$. From these and from Eq. (4) we shall see that the adiabat $p=p(v)$ is more sensitive to the accuracy of evaluation than its $p=p(u)$ version. Let us now consider an impedance equation of the Tait type[3]:

$$\frac{-dp}{dv}=\frac{p+B}{A}, \tag{5}$$

from which we obtain $K=-v\,dp/dv=v(p+B)/A$, $K'=dK/dp=v/A-1$, and therefore $A=v_0/(K_0'+1)$ and $B=K_0/(K_0'+1)$. Here K is the isentropic bulk modulus, and constants A and B become self-explanatory. Also, we have $Z_0=(B/A)^{1/2}$. Straightforward integration of Eqs. (4) and (5) yields insentropes

$$p(u)=(B/A)^{1/2}u+(4A)^{-1}u^2, \tag{6}$$

$$p(v)=B[\exp(v_0-v)/A-1] \tag{7}$$

which correspond to Eqs. (2) and (3), respectively.

Since entropy rise or shock heating is to be accounted for by third-order terms[8] such as of u^3, $(v_0-v)^3$, etc., Eqs. (2) and (6) appear undistinguishable for the Hugoniot adiabat or the Tait isentrope. Such a coincidence leads to the weak-shock equation

$$U=\rho_0^{-1}(B/A)^{1/2}+(4\rho_0A)^{-1}u, \tag{8}$$

which is indeed Eq. (1) with $a_0=\rho_0^{-1}(B/A)^{1/2}\doteq(K_0/\rho_0)^{1/2}$ and $a_1=(4\rho_0A)^{-1}=\frac{1}{4}(K_0'+1)$ as already given consistently in an earlier paper.[9] Here, use of K_0' yields a unique value of a_1 versus two different results of γ_0 from the Slater and Dugdale-MacDonald formulas.[1,9] It should be noted that Eqs. (3) and (7) can be coupled[9,10] in the Grüneisen equation of state to exhibit a second-order contact: $p_H(v_0)=p(v_0)=0$, $-(a_0/v_0)^2=p_H'(v_0)=p'(v_0)=-B/A$, and $4a_1a_0^2/v_0^3=p_H''(v_0)=p''(v_0)=B/A^2$. In other words the Hugoniot curve of Eq. (3) will separate from the isentropic curve of Eq. (7) on the third order—cf. the two equations in binomial expansion. It is of interest to note yet another property, namely, zero impedance. From Eq. (7) we have $dp/dv=0$ at $p=-B=-K_0/(K_0'+1)$. Using Eq. (3) likewise, we get $dp_H/dv=0$ at $p_H=-\rho_0a_0^2/4a_1=-K_0/(K_0'+1)$. The merging of the two adiabatic curves in the negative-pressure region seems to provide further evidence in favor of using the Tait isentrope as a heuristic tool. Qualitatively, the constant B may be regarded as the dynamic strength of solid material associated with spalling into zero impedance. A final remark on the shock adiabat is noteworthy. While Eq. (2) does not contain an entropic term at all, its transform, Eq. (3), can be expanded to give one such as $p_H'''(v_0)(v_0-v)^3/3!$. Without resort to the third derivative of Eq. (3), it is more appropriate to refer to the entropy rise $\Delta S=p_H''(v_0)(v_0-v)^3/12T_0$ with T_0 denoting the initial absolute temperature. For a given specific heat c_v, the shock temperature is approximately $T=T_0\exp(\Delta S)/c_v$.

[1]R. G. McQueen and S. P. Marsh, J. Appl. Phys. 31, 1253 (1960).
[2]L. V. Al'tshuler, B. N. Moiseev, L. V. Popov, G. U. Simakov, and R. F. Trunin, Sov. Phys.-JETP 27, 420 (1968).
[3]R. H. Cole, *Underwater Explosions* (Dover, New York, 1965), pp. 25, 38, 74, and 77.
[4]J. Berger and S. Joigneau, C.R. Acad. Sci. (Paris) 249, 2506 (1959).
[5]A. L. Ruoff, J. Appl. Phys. 38, 4976 (1967).
[6]F. J. Heymann, J. Basic Eng. 90, 400 (1968).
[7]G. W. Swan, J. Phys. D 4, 1077 (1971).
[8]R. Courant and K. O. Friedrichs, *Supersonic Flow and Shock Waves* (Interscience, New York, 1948), pp. 143, 156, 157 and 211.
[9]Y. K. Huang, J. Appl. Phys. 42, 3212 (1971).
[10]Y. K. Huang, J. Appl. Phys. 42, 4084 (1971).

A third-order adiabat for solids at high pressure

Y. K. Huang

Watervliet Arsenal, Watervliet, New York 12189
(Received 23 December 1974; in final form 31 March 1975)

This paper seeks to establish a new isentrope in the form of a generalized Murnaghan equation, by fitting its parameters with compressibility constants which are accurate to the third order Such an isentrope corresponds also closely to the generalized shock adiabat of an earlier investigation by the author. Inasmuch as few third-order compression data and equations are available in the literature, this work is meant to provide one using reliable data and results from shock compression of solids

PACS numbers: 62 50., 62 20 D

In a recent paper[1] we have examined a number of empirical high-pressure equations which are correct only to the second order. So we took a step further to formulate a more general representation which turned out to be simple and accurate, including all the second-order equations as special cases. But our semianalytical treatment[1] is still not clear enough to separate isothermal compressibility from isentropic compressibility. Strictly speaking, these compressibilities differ on all orders of magnitude. In this paper we will only be concerned with isentropic and shock compressions of solids, which can be related in a one-to-one correspondence. Elsewhere[2] we have considered the third-order or generalized shock adiabat in detail. Using such a shock adiabat, we have already expressed the isentrope in the third-order polynomial form.[3] In what follows, it is of interest to consider the nonlinear representation

$$K=K_0(1+p/A)^M, \tag{1}$$

where K is the bulk modulus of isentropic compression at pressure p, and K_0, A and M are appropriate constants. Using $K=-v\,dp/dv$, $K'=dK/dp$, and $K''=d^2K/dp^2$, we immediately obtain

$$M=K_0'^2/(K_0'^2-K_0K_0''), \tag{2}$$

$$A=K_0K_0'/(K_0'^2-K_0K_0''), \tag{3}$$

where subscript 0 refers to initial values at $p=0$. Apparently, exponent M is dimensionless and parameter $A=MK_0/K_0'$ is the pertinent adjustable constant which has the same unit as pressure. Both constants are sensitive to accurate data fitting, and hence Eqs. (1)–(3) will be limited to the third order. Let us rewrite Eq. (1) in the form

$$p=A[(K_0^{-1}K)^{1/M}-1] \tag{4}$$

$$=A\{[1+(1-M)A^{-1}K_0\ln(v_0/v)]^{1/(1-M)}-1\}$$

which may be considered as a generalization of the Murnaghan equation[4] in analogy to our earlier formulation.[1] We will soon show that Eq. (1) or (4) calls for $\frac{1}{2}\leqslant M\leqslant 1$ or $K_0'^2\geqslant -K_0K_0''\geqslant 0$. It is worthwhile to examine the two extreme cases. With $K_0''=0$ in Eq. (2), we get $M=1$ and hence

$$p=(K_0/K_0')(K/K_0-1) \tag{5}$$

$$=(K_0/K_0')[(v_0/v)^{K_0'}-1]$$

by Eqs. (1) and (3). Now Eq. (5) is exactly the original Murnaghan equation[4] which has been used widely in the literature, either as an isotherm or as an isentrope. Note that Eq. (2) also implies $M=1$ as the limiting case with $K_0'^2\gg -K_0K_0''$ (e.g., the Dugdale-MacDonald formula[5] $\gamma_0=2$, $K_0'=5$, $-K_0''\ll 25/K_0$). On the other hand, Eq. (2) gives $M=\frac{1}{2}$ for small K_0', implying very small $-K_0''=K_0'^2/K_0$ (e.g., the Slater formula[6] $\gamma_0=0$, $K_0'=\frac{1}{3}$, $-K_0''=1/9K_0$). In this extreme case, Eqs. (1) and (3) yield

$$p=(K_0/2K_0')[(K/K_0)^2-1]$$

$$=[K_0\ln(v_0/v)][1+\tfrac{1}{2}K_0'\ln(v_0/v)]. \tag{6}$$

Thus, all weakly nonlinear isentropes with $K_0''\lesssim 0$ can be described by Eq. (1) or (4) to the third order with $\frac{1}{2}\leqslant M\leqslant 1$. Such a representation using constants K_0, K_0', and K_0'' turns out to correspond also closely to the generalized shock adiabat[2] based on $U=a_0+a_1u-a_2u^2$ with a_0, a_1, and a_2 all denoting positive constants (here U being the shock velocity and u the particle velocity). See Eqs. (12)–(14).

Further remarks are of relevance here. Although they are not widely used, the following two forms have been proposed[7,8] as a modification of the original Murnaghan expression:

$$p=\alpha[(v_0/v)^\beta-1][(1+\delta)+(1-\delta)(v_0/v)^\beta]^{-1}, \qquad (7)$$

$$p=\theta^{-2}(\theta K_0+\eta)[(v_0/v)^\theta-1]-(\eta/\theta)\ln(v_0/v) \qquad (8)$$

with constants $\beta=(K_0'^2-2K_0K_0'')^{1/2}\neq 0$, $\alpha=2K_0/\beta$, $\delta=K_0'/\beta$, $\theta=K_\infty'$ (as $p,K\to\infty$), and $\eta=K_0(K_0'-K_\infty')$. It can be verified that Eqs. (7) and (8) originate from[7,8]

$$\begin{aligned}K&=K_0+K_0'p+\tfrac{1}{2}K_0''p^2\\&=\tfrac{1}{2}\alpha\beta\left(1+\frac{\delta+1}{\alpha}p\right)\left(1+\frac{\delta-1}{\alpha}p\right)\\&=-v\frac{dp}{dv},\end{aligned} \qquad (9)$$

$$\begin{aligned}K'&=\theta+\frac{\eta}{K}\\&=-\frac{d\ln K}{d\ln v},\end{aligned} \qquad (10)$$

respectively. Clearly, Eq. (7) appears to be suitable for $K_0''<0$ and $\beta>0$. For $K_0''=0$, we get $\beta=K_0'$, $\alpha=2K_0/K_0'$, and $\delta=1$. Now Eqs. (7) and (9) reduce to Eqs. (5) and (1) with $M=1$, respectively. For $\beta=0$ or $K_0'^2=2K_0K_0''$, Eq. (9) leads to

$$\begin{aligned}p&=(2K_0/K_0')[(K_0^{-1}K)^{1/2}-1]\\&=[K_0\ln(v_0/v)][1-\tfrac{1}{2}K_0'\ln(v_0/v)]^{-1}\end{aligned} \qquad (11)$$

which appears rather similar to Eq. (6). But Eq. (11) is indeed a special case of Eq. (1) or (4) with $M=2$ and $K_0''>0$, which we will not include herein. For Eqs. (8) and (10) it seems more appropriate to use $\theta=K_0'+K_0K_0''/K_0'$ and $\eta=-K_0^2K_0''/K_0'$ as below. Thus, Eq. (8) with $\eta=0$ or $K_0''=0$ reduces to Eq. (5) or the original Murnaghan. On the other hand, Eq. (10) with $\theta=0$ or $-K_0''=K_0'^2/K_0$ becomes a special case of Eq. (1) with $M=\frac{1}{2}$. Then Eq. (8) should be replaced by Eq. (6). Moreover, Eq. (10) implies Eq. (9) as follows. Substituting Eq. (9) into Eq. (10), we obtain

$$\begin{aligned}K'&=\theta+\eta(K_0+K_0'p+\tfrac{1}{2}K_0''p^2)^{-1}\\&\approx(\theta+\eta/K_0)-\eta K_0^{-2}K_0'p.\end{aligned}$$

Direct differentiation of Eq. (9) yields

$$K'=K_0'+K_0''p.$$

These two results again lead to $\eta=-K_0^2K_0''/K_0'$ and $\theta=K_0'+K_0K_0''/K_0'$ as mentioned before. It is worth noting that both Eqs. (7) and (8) can be considered as a version of Eq. (4) with $M=2\delta^2/(\delta^2+1)$, $A=\alpha\delta/(\delta^2+1)$, and with $M=(\theta K_0+\eta)/(\theta K_0+2\eta)$, $A=K_0^2/(\theta K_0+2\eta)$, respectively. Thus, Eqs. (1)–(4) are of considerable simplicity and generality. As far as Eq. (1) is concerned, we need only to consider $\frac{1}{2}\leqslant M\leqslant 1$ with $K_0''\lesssim 0$ as dictated by the sluggish variation of the Grüneisen γ with most solids.[2]

Since we have accurate interrelations[3] based on a third-order shock adiabat,[2] it is expedient to fit Eqs. (1)–(4) with shock-compression data. Let us take the results[3]

$$K_0=a_0^2/v_0, \qquad (12)$$

$$K_0'=4a_1-1, \qquad (13)$$

$$-K_0K_0''=2a_1(\gamma_0-a_1+2)+12a_0a_2, \qquad (14)$$

where a_0, a_1, and a_2 are all shock-compression properties, and K_0, K_0', and K_0'' are not isothermal but isentropic compressibility constants. Substituting Eqs. (13) and (14) in Eq. (2), we get the general expression

$$M=(4a_1-1)^2/[(4a_1-1)^2+2a_1(\gamma_0-a_1+2)+12a_0a_2] \qquad (15)$$

to which both the Dugdale-MacDonald and Slater formulas are applicable. Once M is accurately determined, we simply get $A=MK_0/K_0'$. To illustrate, let us take copper with $a_0=3.96$ km/sec, $a_1=1.5$, $a_2=0$, $K_0'=5$, and $\gamma_0=2$ as a typical example to the second order.[9] Using the Dugdale-MacDonald formula, we get $v_0\gamma_0'=2.30$ which is not too far off the Los Alamos approximation[9] $v_0\gamma_0'=\gamma_0=2$. From these and Eqs. (13)–(15) we get $-K_0K_0''=7.50$ and $M=0.77$. It is obvious that we should get $-K_0K_0''>7.50$ with $a_2\gtrsim 0$. As copper has been given[10] $a_0=3.91$ km/sec, $a_1=1.55$, and $a_2=0.0144$ sec/km to the third order, so we deduce $v_0\gamma_0'=2.79$, $-K_0K_0''=8.59$, and $M=0.76$. If we used the Tait value[11] $-K_0K_0''=K_0'+1=6$ (too small), then we should have obtained $M=0.81$ (too large). If we used the Rice value[12] $v_0\gamma_0'=\gamma_0(\gamma_0+1)=6$ (too large), then we should have obtained $-K_0K_0''=2v_0\gamma_0'+\frac{2}{3}K_0'-\frac{4}{9}=14.89$ (too large) and $M=0.62$ (too small). These considerations indicate that our model is adequately close for both the second-order and the third-order approximations. Since accurate data for a_2, K_0'', and γ_0' are still lacking, we have exercised extra caution to evaluate or use these constants adequately. In conclusion this paper has the motivation to formulate a new isentrope which turns out to be not only relatively simple but also an order more accurate than the well-known Tait and Murnaghan equations for solids at high pressure.

[1]Y.K. Huang and C.Y. Chow, J. Phys. D 7, 2021 (1974).
[2]Y.K. Huang, J. Appl. Phys. 42, 3212 (1971).
[3]Y.K. Huang, J Appl. Phys. 42, 4084 (1971).
[4]F.D. Murnaghan, Proc. Natl. Acad. Sci. USA 30, 244 (1944).
[5]J.S. Dugdale and D.K.C. MacDonald, Phys. Rev. 89, 832 (1953).
[6]J.C. Slater, *Introduction to Chemical Physics* (McGraw-Hill, New York, 1939), p. 239.
[7]P.S. Ho and A.L. Ruoff, J. Phys. Chem. Solids 29, 2101 (1968).
[8]A. Keane, Aust. J. Phys. 7, 322 (1954).
[9]R.G. McQueen and S.P. Marsh, J. Appl. Phys. 31, 1253 (1960).
[10]D.J. Pastine and D. Piacesi, J. Phys. Chem. Solids 27, 1783 (1966).
[11]Y.K. Huang, J. Appl. Phys. 45, 2346 (1974).
[12]M.H. Rice, J. Phys. Chem. Solids 26, 483 (1965).

J. Phys. D: Appl. Phys., Vol. 7, 1974. Printed in Great Britain. © 1974

The generalized compressibility equation of Tait for dense matter

Y K Huang† and C Y Chow‡
† 16 Verdi Boulevard, Latham, New York 12110, USA
‡ General Electric Company, Schenectady, New York 12345, USA

Received 19 April 1974, in final form 30 July 1974

Abstract. In this paper the compressibility equation of Tait is generalized in a basic form with three arbitrary parameters. Data-fitting of these parameters can be made correct to the third order. Thus, the generalized Tait yields a more flexible equation of state that is also related simply to several other well-known equations of high pressure, all accurate only to the second order.

1. Generalization of Tait's equation of state

A number of well-known empirical equations of state (listed below) all satisfy an equation of the form

$$K/v = -p' = Z^2 = Z_0^2\,(1+p/B)^N \tag{1}$$

where K is the bulk modulus of compression, v the specific volume, p the pressure, p' the abbreviation of dp/dv, Z the acoustic impedance, and Z_0, B, and N three arbitrary parameters which can be fitted with appropriate constants.

Tait's equation (Hayward 1967) is

$$v_0 p/(v_0 - v) = (p + B_0)/A_0 \tag{2}$$

giving

$$N=2,\ B=B_0,\ \text{and}\ Z_0^2 = B_0/A_0 v_0.$$

Tammann's equation (Hayward 1967) is

$$p = B[\exp (v_0 - v)/A - 1] \tag{3}$$

giving

$$N=1\ \text{and}\ Z_0^2 = B/A.$$

Bridgman's (1949) isotherm is

$$\Delta v/v_0 = 1 - v/v_0 = ap - bp^2 \tag{4}$$

implying

$$N=-1,\ B=-a/2b,\ \text{and}\ Z_0^2 = (av_0)^{-1}.$$

Murnaghan's equation (Cook and Rogers 1963, Rodionov 1970) is

$$p = B[(v_0/v)^{K_0'} - 1] \tag{5}$$

implying

$$v_0 = B/Z_0^2\,(N-1).$$

This condition is readily verifiable with $N=1+K_0'^{-1}$ and $Z_0^2=BK_0'/v_0$. It is worth noting that we can also write $B=K_0/K_0'$. Let us consider a Hookean solid with its isotherm given by

$$p=K_0(1-v/v_0) \tag{6}$$

which implies

$$N=0,\ B=-K_0, \text{ and } Z_0^2=K_0/v_0.$$

Obviously, equation (6) can be given by equation (5) with $K_0'=-1$.

Except for the Hookean equation, all these equations of state have three arbitrary parameters and so can be fitted to the second order. By integrating equation (1), we obtain a more general equation of state with four parameters:

$$v_0-v=B(N-1)^{-1}Z_0^{-2}\,[1-(1+p/B)^{1-N}] \tag{7}$$

where v_0 is the initial volume at $p=0$. Equation (7) may be fitted correct to the third order, with

$$Z_0^2=K_0/v_0=-p_0' \tag{8}$$

$$B=[K_0'/K_0-K_0''/(K_0'+1)]^{-1}=p_0'(2p_0''/p_0'-p_0'''/p_0'')^{-1} \tag{9}$$

$$N=(K_0'+1)\,[K_0'-K_0K_0''/(K_0'+1)]^{-1}=(2-p_0'p_0'''/p_0''^2)^{-1}. \tag{10}$$

Here we have used

$$\left.\begin{aligned} &K=-vp',\ K'=\mathrm{d}K/\mathrm{d}p=-\mathrm{d}\,(\ln K)/\mathrm{d}\,(\ln v)=-1-vp''/p',\\ &\text{and}\\ &K''=\mathrm{d}^2K/\mathrm{d}p^2=-[p'p''+v\,(p'p'''-p''^2)]/p'^3 \end{aligned}\right\}\ \text{at } p=0.$$

As shown by equations (9) and (10), the accuracy of constants B and N depends rather upon the accurate values of higher-order compression properties. At present there is still a lack of accurate data for K_0'' and p_0''', although we may sometimes calculate these using the Mie potential (Grüneisen 1959) or shock-compression formulae (Huang 1971). But such calculated results are rough approximation only. Returning to equation (1), we have $p=-B$ at $Z=0$. This seems to indicate that constant B has a bearing upon the spall resistance of solid material (Huang 1974).

2. Discussion

For a demonstration of our generalized model other than those considered before, it is essential to evaluate the principal constants Z_0, B and N for reasonable accuracy. Let us use Mie's equation of state (Grüneisen 1959):

$$p=3K_0\,(n-m)^{-1}\,[(v_0/v)^{1+n/3}-(v_0/v)^{1+m/3}] \tag{11}$$

with positive integer $n>m$. From equation (11), we deduce

$$\left.\begin{aligned} p_0' &= -K_0/v_0 < 0 \\ p_0'' &= (n+m+9)\,K_0/3v_0^2 > 0 \\ p_0''' &= -K_0 v_0^{-3}\,[(n^2+nm+m^2)/9+2(n+m)+11] < 0 \\ K_0' &= (n+m+6)/3 > 0 \\ K_0'' &= -(n+3)\,(m+3)/9K_0 < 0 \end{aligned}\right\} \qquad (12)$$

which depict the compression behaviour and properties of solids qualitatively (Huang 1971). It is of interest to consider three special cases of equations (11) and (12), with the pertinent p_0''' and K_0'' as a rough estimate of the higher-order compression properties. Thus, we have the Bardeen (1938) model for alkali metals with $m=1$, $n=2$, $K_0'=3$, and $K_0''=-2{\cdot}22/K_0$; the Birch (1952) model for simple solids with $m=2$, $n=4$, $K_0'=4$, and $K_0''=-3{\cdot}89/K_0$; and the Yayanos (1970) model for sea water with $m=6$, $n=9$, $K_0'=7$, and $K_0''=-12/K_0$. Using these values in conjunction with equations (9) and (10), we get $N=1{\cdot}13$ and $B=K_0/3{\cdot}56$ for the Bardeen model, $N=1{\cdot}05$ and $B=K_0/4{\cdot}78$ for the Birch model, and $N=0{\cdot}94$ and $B=K_0/8{\cdot}50$ for the Yayanos model. Since all these investigators have provided considerable evidence for their equations of state, it is worth noting that these examples in transform with $N\approx 1$ and $B\approx K_0/(K_0'+1)$ are reasonably comparable to a recent verification based on shock compression (Huang 1974).

Using shock-compression data and formulae (Huang 1971), we can likewise calculate K_0', K_0'', and hence N and B by equations (9) and (10). In other words, from shock compressibility we can also deduce isentropes or isotherms in the form of equation (7). Experimentally, our basic equation (1) can be tested or fitted with ultrasonic data. On the other hand, equations (8)–(10) can be used for evaluating higher-order compression properties (p_0''', K_0'') in terms of given Z_0, B, and N.

Acknowledgment

The authors are grateful to the referees for their comments and suggestions to revise this manuscript for clarity and improvement.

References

Bardeen J 1938 *J. Chem. Phys.* **6** 372–8
Birch F 1952 *J. Geophys. Res.* **57** 227–86
Bridgman P W 1949 *The Physics of High Pressure* (London: Bell and Sons)
Cook M A and Rogers L A 1963 *J. Appl. Phys.* **34** 2330–6
Grüneisen E 1959 *The State of A Solid Body* (Washington, DC: US Government Printing Office) NASA RE-2-18-59W
Hayward A T J 1967 *Br. J. Appl. Phys.* **18** 965–77
Huang Y K 1971 *J. Appl. Phys.* **42** 3212–5
—— 1974 *J. Appl. Phys.* **45** 2346–7
Rodionov K P 1970 *Physics Metals Metallogr., N.Y.* **29, 6** 51–9
Yayanos A A 1970 *J. Appl. Phys.* **41** 2259–60

Reprinted from:

September 1972

A Special Class of Ideal Quantum Gases

Y. K. HUANG
Benet Research and Engineering Laboratories
Watervliet Arsenal,
Watervliet, New York 12189
(Received 8 November 1971; revised 25 February 1972)

This essay is motivated to add a useful note to the contemporary literature of ideal gases. For brevity, a new description is given for a special class of ideal quantum gases (the "super ideal gases") which are characterized here by two compact equations of state. Within such a framework all fundamental properties of these gases are readily expressed in terms of two constants known as the state indices. In order to test our results of deduction, six typical examples are given with comparable numerical values. While all the examples must otherwise have resort to statistical calculations, our evaluation for them turns out to be straightforward as well as heuristic.

INTRODUCTION

Degenerate gases have long been an interesting and useful topic of statistical physics.[1,2] From the thermodynamic point of view, the simpler method and results of ideal classical gases can be so generalized as to cover a number of degenerate gases. As such, Landsberg[3,4] has given an elegant summary of the general behavior and properties of ideal quantum gases in terms of a parameter and an undetermined function.[3] The latter two quantities stem from his use of only one equation of state [cf., our Eq. (5)] for the characterization of ideal quantum gases, and their ultimate evaluation must have resort to statistical mechanics. We consider that statistical calculation can be eschewed as a separate effort by itself. Thus, we take a new approach by introducing an additional equation of state [viz. Eq. (6)] as our basic tool in this paper. Now this final class of ideal quantum gases can be called "super ideal," and characterized by two dimensionless parameters known as the "state indices" (g, j). Also, we will express all its fundamental properties as logarithmic derivatives which can be put again in terms of the two state indices. Since these state indices will only take particular (also deducible) values as our input data for a special class of ideal quantum gases (nonrelativistic), we seek to confine our attention to six typical examples as a demonstration of our methodology. It turns out that our explicit results are in viable form and complementary to Landsberg's general scheme.

EQUATIONS OF STATE

Let the five state variables of a simple system be denoted as follows: P, equilibrium pressure; V, specific volume; T, absolute temperature; S, specific entropy; and U, specific internal energy. All lower-case subscripts will refer to the corresponding capital variables that are held constant during partial differentiation. Then the basic equation of thermodynamics is given by

$$dU = TdS - PdV, \tag{1}$$

which can serve to specify these equations of state:

$$U = U(V, S), \tag{2}$$

$$P = -(\partial U/\partial V)_s, \tag{3}$$

$$T = (\partial U/\partial S)_v. \tag{4}$$

Equations (3) and (4) appear to suggest that we may relate the three energy functions (PV, ST, U) to the two state indices (g, j) as follows:

$$g = PV/U = -(\partial \ln U/\partial \ln V)_s, \tag{5}$$

$$j = ST/U = (\partial \ln U/\partial \ln S)_v, \tag{6}$$

which may also be regarded as the formal definitions of g and j, respectively. Landsberg[4]

TABLE I. A summary of fundamental properties.

Notation	Remark
$g = PV/U, \quad j = ST/U$	Equations of state with constant indices (g, j)
$U = C_1 V^{-g} S^j$	
$F = C_2 V^{g/(j-1)} T^{j/(j-1)}$	
$H = C_3 P^{g/(g+1)} S^{j/(g+1)}$	
$G = C_4 P^{-g/(j-g-1)} T^{j/(j-g-1)}$	
$\alpha T = j/(j-g-1)$	
$c_p/S = (g+1)/(j-g-1), \quad c_p = \alpha H$	
$c_v/S = 1/(j-1)$	
$B_s/P = g+1, \quad (PV^{g+1})_s = \text{constant}$	
$g = \Gamma, \quad (TV^g)_s = \text{constant}$	
$B_t/P = (j-g-1)/(j-1),$	
$[PV^{(j-g-1)/(j-1)}]_t = \text{constant}$	
$\gamma = c_p/c_v = B_s/B_t$	
$\quad = (g+1)(j-1)/(j-g-1)$	
$(\partial U/\partial V)_t = P/(j-1)$	Ideal classical gases with $j \to \infty$
$(\partial T/\partial P)_h = V/S$	

has shown that the state index g is identical with the Grüneisen constant Γ for an ideal quantum gas. In addition, let us consider j a certain constant to be deducible in one way or the other. So we seek to confine our attention to a special class of ideal quantum gases for which both g and j are constants. Now Eqs. (5) and (6) can be integrated to yield

$$U = C_1 V^{-g} S^j \tag{7}$$

as an explicit form of Eq. (2), C_1 being an appropriate constant. By definition and through a few manipulations with Eqs. (5)–(7), we obtain

$$F = (1-j)\,U = C_2 V^{g/(j-1)} T^{j/(j-1)}, \tag{8}$$

$$H = (1+g)\,U = C_3 P^{g/(g+1)} S^{j/(g+1)}, \tag{9}$$

$$G = (1-j+g)\,U = C_4 P^{-g/(j-g-1)} T^{j/(j-g-1)}, \tag{10}$$

where F is the Helmholtz free energy, H is the specific enthalpy, G is the Gibbs free enthalpy, and C_2, C_3, and C_4 are all constants. Equations (7)–(10) are the four canonical forms from each of which can follow a complete description of the equilibrium state for the special ideal gas in terms of the pertinent thermodynamic coordinates and/or by use of the Maxwell relations.[3]

FUNDAMENTAL PROPERTIES

For a simple description of the macroscopic system under consideration we will use its five measurable properties: α, thermal coefficient of volumetric expansion; c_p, specific heat at constant

TABLE II. A summary of numerical examples.

Brief description

1. Completely degenerate electron gas at 0°K (cf. Ref. 1, p. 161):
   $g = 2/3, \quad j = 0$
   $F_c = U_c = AV^{-2/3}$ (Fermi energy, A being constant)
   $G_c = H_c = 5U_c/3$
   $S = 0, \quad c_p = c_v = 0, \quad \alpha T = 0$
   $(PV^{5/3})_s = P_c V^{5/3} = \text{constant}$

2. Free electron gas below degeneracy temperaturer (cf. Ref. 1, p. 165; Ref. 3, p. 208):
   $g = 2/3, \quad j = 2$
   $F_* = -U_* = 3G_* = -3H_*/5$
   $c_v = S_*, \quad c_p = 5S_*, \quad \gamma = 5, \quad \alpha T = 6$
   $(PV^{5/3})_s = \text{Constant}, \quad (TV^{2/3})_s = \text{Constant}$
   $(PV^{1/3})_t = \text{Constant}$ as verifiable with $F_* = -AV^{2/3}T^2$

3. Bose degenerate gas below critical temperature (cf. Ref. 1, p. 170; Ref. 3, p.211)
   $g = 2/3, \quad j = 5/3$
   $G_* = 0, \quad H_* = S_*T = 5U_*/3 = -5F_*/2, \quad P_* = P_*(T)$
   $c_v = 3S_*/2, \quad \alpha B_t = S_*/V, \quad B_t = 0, \quad \alpha \to \infty, \quad c_p \to \infty$
   $(PV^{5/3})_s = \text{Constant}, \quad (TV^{2/3})_s = \text{Constant}$

4. Photon gas or black-body radiation (cf. Ref. 1, p. 175; Ref. 3, p. 211)
   $g = 1/3, \quad j = 4/3$
   $G = 0, \quad F = -ST/4 = -U/3 = -H/4, \quad P = P(T)$
   $c_v = 3S, \quad \alpha B_t = S/V, \quad B_t = 0, \quad \alpha \to \infty, \quad c_p \to \infty$
   $(PV^{4/3})_s = \text{constant}, \quad (TV^{1/3})_s = \text{constant}$

5. Phonon gas or quantized elastic waves in Debye solids (cf. Ref. 1, p. 184; Ref. 3. p. 211)
   $g = 1/3, \quad j = 4/3$
   $G_* = 0, \quad H_* = S_*T = -4F_*$
   $c_v = 3S_*, \quad \gamma \approx 1, \quad \alpha T \approx 0$
   $(PV^{4/3})_s = \text{Constant}, \ (TV^{1/3})_s = \text{Constant}$

6. Ideal classical gases (cf. Ref. 1, p. 125; Ref. 3, p. 207)
   $g = \gamma - 1, \quad j \to \infty \ (S \gg c_v)$
   $\gamma = 5/3,\ 7/5$ or $4/3$
   $(\partial U/\partial T)_v = 0, \quad (\partial T/\partial P)_h = 0, \quad \alpha T = 1$
   $PV = \text{constant}$ for isotherm
   $(PV^\gamma)_s = \text{constant}, \quad (TV^{\gamma-1})_s = \text{constant}$

[a] It should be noted that conduction electrons in metals under high pressure[5] behave more or less like this ideal gas. Isotope helium three (a quantum liquid at very low temperature) may also be described by this model, in virtue of its Fermi–Dirac distribution.

pressure; c_v, specific heat at constant volume; B_s, bulk modulus of isentropic compression; and B_t, bulk modulus of isothermal compression. Let Γ denote the Grüneisen ratio[3,4] and $\gamma = c_p/c_v$. We may now summarize all these fundamental properties in the following dimensionless forms:

$$\alpha T = (\partial \ln V/\partial \ln T)_p, \tag{11}$$

$$c_p/S = (\partial \ln S/\partial \ln T)_p, \tag{12}$$

$$c_v/S = (\partial \ln S/\partial \ln T)_v, \tag{13}$$

$$B_s/P = -(\partial \ln P/\partial \ln V)_s, \tag{14}$$

$$B_t/P = -(\partial \ln P/\partial \ln V)_t, \tag{15}$$

$$\gamma = c_p/c_v = B_s/B_t = 1 + \Gamma\alpha T, \tag{16}$$

$$\Gamma = \alpha V B_t/c_v = V(\partial P/\partial U)_v = -(\partial \ln T/\partial \ln V)_s, \tag{17}$$

In Table I we have further expressed these quantities all in terms of g and j. There we note $g = \Gamma$ and $j = 1 + S/c_v$. It can be verified that statistical calculation yields S/c_v, and hence j, a constant for several degenerate gases.[1] Also, it should be noted that the empirical laws of Boyle, Joule, and Kelvin are dispensable[2,3] for ideal quantum gases. The last two coefficients of Table I reduce to zero for ideal classical gases with $S \gg c_v$.

EXAMPLES

Strictly speaking, the numerical values of g and j should be sought via statistical calculations.[1,3] As mentioned earlier, the Grüneisen constant Γ, and hence g, is a certain fraction for a given degenerate gas. To form a special class, some degenerate gases are characterized by $j = g + 1$ as a result of $G = 0$. Such are the examples as photons, phonons, and degenerate bosons. The peculiar behavior of these degenerate gases may also be described by $c_v = S/g$, $\alpha B_t = S/V$, $B_t = 0$, $\alpha \to \infty$, and $c_p \to \infty$ as attributable to their thermic equation of state $P = P(T)$ or $P_* = P_*(T)$. Here it should be remarked that the subscript $*$ refers to the reduced properties; e.g.,

$$P_* = P - P_c = -(\partial F_*/\partial V)_t,$$

$$S_* = S = -(\partial F/\partial T)_v,$$

$$U_* = U - U_c = F_* + S_* T,$$

with zero-point values denoted by the subscript c meaning cold. In Table II we give six typical examples whose property values can be verified readily with Refs. 1 and 3.

CONCLUSION

Only two state indices (g, j) are needed for the complete thermodynamic description of a special class of ideal quantum gases, including the classical Boltzmann gas as a special case. The constant g has already been identified as the Grüneisen Γ, and the other index is given either by $j = 1 + S/c_v$ or by $j = g + 1$. In essence, the equations of state are strikingly simple for such a group of ideal gases.

ACKNOWLEDGMENTS

Many thanks are due to the reviewer and editors without whose helpful criticism and suggestions it would not be possible to issue this paper in its present form. Also, moral support has always been given by my directors for research effort like this work.

[1] L. D. Landau and E. M. Lifshitz, *Statistical Physics* (Pergamon, London, 1958), pp. 51, 125, 158–161, 165, 170, 175, and 184.

[2] H. Einbinder, Phys. Rev. **74**, 805 (1948).

[3] P. T. Landsberg, *Thermodynamics with Quantum Statistical Illustrations* (Wiley, New York, 1961), pp. 41, 71, 72, 194, 201–204, 207, 208, 210, and 211.

[4] P. T. Landsberg, Amer. J. Phys. **29**, 695 (1961).

[5] Y. K. Huang, J. Chem. Phys. **53**, 571 (1970).

Foundations of Physics, Vol. 4, No. 2, 1974

Analytical Model for Super-ideal Gases

Y. K. Huang

Watervliet Arsenal, Watervliet, New York 12189

Received June 29, 1973

The equations of state for the ideal classical gas are generalized with two characteristic constants known as the state indices. A simple and complete representation is developed for the super-ideal gas, with explicit results which are general enough to cover a wide range of equilibrium systems and states. Evidence and justification are provided in terms of examples and results of statistical physics. The new model should prove useful for treating problems involving physical behavior and properties which might otherwise call for specialized or advanced methods of statistical mechanics.

1. INTRODUCTION

The classical model of ideal gases has been a heuristic device for solutions of many problems in thermodynamics, interior ballistics, gas dynamics (supersonics, explosions, star formation), and thermal physics. But its validity is rather restricted to the extreme condition of low density, low pressure, and high temperature. Using the simple equations of state of this model as a basis, we find it justifiable to generalize them into a super-ideal model which can represent a variety of ordinary and degenerate gases consistently. Thus, the new model is much simpler and more general than the various statistical models which it can represent. Although it is somewhat abstract in form, the super-ideal model can provide explicit results over a wide range of equilibrium states as governed by the first, second, and third laws of thermodynamics. Such results suffice to describe the ideal behavior and properties of many systems, which are verifiable by laborious calculations of statistical mechanics. In this sense our new model appears to be more versatile than those of classical and quantum gases. It is the new method and results which we emphasize in this and an earlier paper.

2. IDEAL MODEL

It is well known that the ideal model of classical gases can be described completely by the two equations of state

$$PV = RT \tag{1}$$

$$E = c_v T \tag{2}$$

where P denotes the equilibrium pressure, V the molar volume, T the absolute temperature, E the internal energy per mole, R the universal gas constant, and c_v the molar heat capacity at constant volume. These equations hold good only for states of gases with low density, at low pressure, and at high temperature so that the intermolecular attraction need not be considered, on the assumption of constant c_v ($E \propto T$). It is of analytical interest to combine Eqs. (1) and (2) into

$$PV/E = R/c_v = g \tag{3}$$

where g is the adiabatic index, as will soon be indicated clearly by Eqs. (5) and (6). Let us now introduce the basic equation for a simple system in thermodynamic equilibrium:

$$T\,dS = dE + P\,dV \tag{4}$$

where S is the entropy per mole. Eliminating E from Eqs. (3) and (4) and then integrating with $dS = 0$, we get

$$(PV^{1+g})_s = \text{const} \tag{5}$$

Eliminating P from Eqs. (1) and (5), we get

$$(TV^g)_s = \text{const} \tag{6}$$

Both Eqs. (5) and (6) represent the isentrope, reflecting the adiabatic nature of g. Also, Eq. (6) shows that the adiabatic index is identical with the

Grüneisen constant $\gamma = -(\partial \ln T/\partial \ln V)_s = g$. Substituting Eq. (2) in Eq. (4) and integrating for S, we get

$$ST/E = S/c_v = \ln(TV^g) \tag{7}$$

where we have omitted the constant of integration for the sake of simplicity. Consideration of both high temperature and low density indicates that Eq. (7) tends to infinity for the ideal gas.

Other useful properties of the ideal classical gas can also be deduced from Eqs. (1), (2), and (4). But we wish to base our central concepts upon Eqs. (3) and (7). In a recent paper[1] we generalized the above model into the super-ideal model using both Eqs. (3) and (7) in the form

$$PV/E = g \tag{8}$$

$$ST/E = j \tag{9}$$

with constants (g, j) known as the state indices. From Eq. (9) we immediately deduce

$$j = 1 + S/c_v \tag{10}$$

$$E/T = c_v(j-1)/j \tag{11}$$

As $j \to \infty$, Eq. (9) becomes identical with Eq. (10), as shown by Eq. (7). Also, Eq. (11) then assumes the same form as Eq. (2). It is important to note that both Eqs. (9) and (10) are justifiable from the constancy[2] of S/c_v for a number of degenerate gases. In what follows we shall, however, take a new departure from our earlier paper[1] in order to gain deeper insight into the super-ideal model.

3. ANALYTICAL TREATMENT

The canonical representation with $E = E(V, S)$ has already been given in Ref. 1. Here we want to adopt another useful representation with $E = E(V, T)$ by introducing two new indices (ξ, η). Such a step turns out to dispense with the Legendre transform of $E(V, S)$ into $F(V, T)$, the Helmholtz free energy function. Thus, we may write

$$E = CV^{-g}S^j = AV^{\xi}T^{\eta} \tag{12}$$

with

$$A = C^{1/(1-j)}j^{j/(1-j)}, \qquad C \text{ a constant} \tag{13}$$

$$\xi = g/(j-1) \tag{14}$$

$$\eta = j/(j-1) \tag{15}$$

Consequently, we have (see Ref. 1)

$$P = gAV^{\xi-1}T^{\eta} \tag{16}$$
$$S = jAV^{\xi}T^{\eta-1} \tag{17}$$
$$(PV^{1-\xi})_t = \text{const} \tag{18}$$
$$(SV^{-\xi})_t = \text{const} \tag{19}$$
$$(K/P)_t = -(\partial \ln P/\partial \ln V)_t = 1 - \xi \tag{20}$$
$$c_v/S = (\partial \ln S/\partial \ln T)_v = \eta - 1 \tag{21}$$
$$\gamma = -(\partial \ln T/\partial \ln V)_s = \xi/(\eta - 1) \tag{22}$$
$$\alpha T = (\partial \ln V/\partial \ln T)_p = \eta/(1 - \xi) \tag{23}$$
$$c_p/c_v = K_s/K_t = 1 + \gamma\alpha T = 1 + \xi\eta/(1 - \xi)(\eta - 1) \tag{24}$$
$$c_p/S = (1 + \gamma\alpha T)c_v/S = (\xi + \eta - 1)/(1 - \xi) \tag{25}$$
$$(K/P)_s = (1 + \gamma\alpha T)(K/P)_t = (\xi + \eta - 1)/(\eta - 1) \tag{26}$$

where α denotes the volumetric coefficient of thermal expansion, K the bulk modulus of compression, c_p the molar heat capacity at constant pressure, and the lowercase subscripts correspond to the capital symbol held constant during partial differentiation. With Eqs. (14) and (15) substituted in the above, it is easy to compare these results with those listed in Table I of Ref. 1.

From Eqs. (4) and (9) we get

$$S\,dT + P\,dV = (j - 1)\,dE$$

which yields

$$(\partial E/\partial T)_v = S/(j - 1) \tag{27}$$
$$(\partial E/\partial V)_t = P/(j - 1) \tag{28}$$

A glance at Eqs. (8), (9), (12), (27), and (28) shows

$$\xi = (\partial \ln E/\partial \ln V)_t = PV/E(j - 1) = g/(j - 1)$$
$$\eta = (\partial \ln E/\partial \ln T)_v = ST/E(j - 1) = j/(j - 1)$$

These formulas have already been given as Eqs. (14) and (15) through an implicit transformation. Of course, we may also write $\xi = (\partial \ln S/\partial \ln V)_t$ using Eq. (17) or Eqs. (12) and (9). Similarly, using Eq. (16), we may write $\eta = (\partial \ln P/\partial \ln T)_v$. It should be remarked that ξ is the isothermal index. See Eqs. (18) and (19). The other index η stems from the parameter c_vT/E, which is a constant for many degenerate gases.[2]

4. DISCUSSION

Since the state indices (ξ, η) are not independent of (g, j), it suffices to consider the latter for the range of validity. From the Clausius virial[3] we have $3PV = 2E_k + E_p$, where E_k is the kinetic energy and E_p is the potential energy. Substituting the virial in Eq. (8), we get

$$g = (2E_k + E_p)/3E \tag{29}$$

Obviously, we can assert $g \neq 0$. For weakly interacting systems we may put $E_p \ll E_k$ and hence $g = 2/3$ for $E = E_k$ (nonrelativistic kinetic energy) and $g = 1/3$ for $E = 2E_k$ (relativistic). As mentioned in connection with Eq. (6), g is identical with γ. For ideal classical gases with $\alpha T = 1$ [from Eqs. (23) and (1)], we have $g = \gamma = c_p/c_v - 1 = 2/f$ from Eq. (24). Here f is the number of degrees of freedom ($f = 3$, 5, 6 for monatomic, diatomic, and polyatomic gases, respectively). Later we shall also consider $g = 2/3$ and $g = 1/2$ for the limiting states of solids under strong shock compression. These considerations and Eq. (29) appear to suggest that

$$0 < g \leqslant 2/3 \tag{30}$$

The significance of j can be sought from Eqs. (9) and (10). It is interesting to note that these equations require $j = 0$, $S = -c_v = 0$ as $T \to 0^0$K. Then Eq. (24) gives $c_p = c_v$ and $K_s = K_t$. In other words, the third law of thermodynamics is satisfied by Eq. (9). As referred to earlier, j tends to infinity for systems with small heat capacity $c_v \ll S$. See Eq. (10). Therefore we have a wide range of values for j

$$0 \leqslant j \leqslant \infty \tag{31}$$

Without resort to statistical calculation, we shall evaluate (g, j) and hence (ξ, η) from analytical consideration only. The following special cases are worth noting.

First, it is interesting to consider Eq. (12) with $j \to \infty$ and hence $\eta = 1$ from Eq. (15). From Eqs. (14) and (22), g can assume a certain limit, providing $\xi = 0$ (so $g = 0 \times \infty$, $\gamma = 0/0$). Then Eqs. (23) and (24) give $\alpha T = 1$ and $g = c_p/c_v - 1$, respectively. Both Eqs. (21) and (25) show small heat capacities relative to the entropy. All these are the main characteristics of the ideal classical gas. Moreover, Eqs. (18) and (28) express the laws of Boyle and Joule. So we can clearly identify the special case with $(g, j) = (2/f, \infty)$ and $(\xi, \eta) = (0, 1)$, implying $f = 3$, 5, or 6.

Next, let us consider Eq. (16) with $\xi = 1$. That is, the pressure depends upon the temperature only. Substituting this value in Eqs. (14) and (15), we get $j = 1 + g$ and $\eta = 1 + g^{-1}$, respectively. Now Eq. (20) gives $K_t = 0$;

Eq. (21) yields $c_v = S/g$; and Eqs. (23)–(25) show $\alpha T \rightarrow \infty$ and $c_p \rightarrow \infty$. For this group of super-ideal gases the essential index is $g = S/c_v$ or $g = \gamma$. Thus, we have $(\xi, \eta) = (1, 4)$ and hence $(g, j) = (1/3, 4/3)$ for photons and phonons —one at high temperatures and the other at very low temperatures. Bosons may be described with $(g, j) = (2/3, 5/3)$ and $(\xi, \eta) = (1, 5/2)$. See Ref. 1 for more property values.

If we put $\eta = j = 0$ and hence $\xi = -g$ in Eq. (12), then we deduce $c_p = c_v = -S = 0$ from Eqs. (21) and (25). Consideration of these and Eq. (16) leads to $\xi = -2/3$ for a completely degenerate electron gas at 0°K. On the other hand, the free electron gas has $S/c_v = 1$ and $g = 2/3$. So the different characteristic is $j = 2$. See also Ref. 1.

If we put $j = g$ in Eqs. (14) and (15), we have $\xi = \eta$. Substituting these in Eqs. (22)–(24), we get $\alpha T = -g$ and $c_p/c_v = 1 - g^2$. The latter two results ($\alpha < 0$, $c_p < c_v$) are very unusual! We find the peculiar example of a hot plasma with $j = g = 1/3$ (and hence $\xi = \eta = -1/2$). According to the Debye–Hückel theory[4] of completely ionized, rarefied gases at high temperatures, the simple electrostatic or Coulomb attraction is responsible for $\alpha T = -1/3$ and $c_p = 8c_v/9$. While interatomic attraction is absent in all the above examples, the repulsive potential of this example has been neglected.

Another interesting example is furnished by a limiting state of a solid under shock compression—all its nuclei being regarded as harmonic oscillators. According to Debye's theory, we should have[3,5] $K' = dK/dP = 2\gamma + \frac{1}{3}$. A glance at Eqs. (14), (15), (20), and (22) shows that we can get $K' = 1 + \gamma$ by putting $j = \eta = 0$. Now we deduce $1 + g = 2g + \frac{1}{3}$ and hence $g = \frac{2}{3} = -\xi$. Then Eq. (16) may be looked upon as a degenerate form of the Bardeen or Murnaghan equation.[3] On the other hand, the same set of index values associated with the Birch equation[3] would describe the limiting state of a solid as a completely degenerate electron gas, except for a negative pressure component. With $g = \frac{2}{3}$ the maximum shock compression[5] is predicted to be $1 + 2/g = 4$.

The Thomas–Fermi theory[3] of atoms provides yet another description of dense matter at high pressures. In this case the Grüneisen limit[5,6] is $g = \frac{1}{2}$, corresponding to a maximum shock compression of five. Since electrons have $S/c_v = 1$ and hence $j = 2$, we can now describe the limiting state of a solid with $(g, j) = (\frac{1}{2}, 2) = (\xi, \eta)$. Some important properties of this model can be deduced readily from Eqs. (12)–(26).

5. CONCLUSION

In this investigation we have carefully employed both inductive and deductive methods to formulate our analytical model for super-ideal gases.

A simple and complete representation is developed with explicit results which are general enough to cover a variety of special cases, including the classical (Boltzmann) gas. Evidence and justification are provided for our basic formulation, and illustrating examples are given which yield good agreement with numerical data. Analytical results of this investigation should prove useful for treating problems involving physical behavior and properties which might otherwise call for specialized or advanced methods of statistical mechanics.

With considerable emphasis on the analytical basis, this paper was intended to provide a more refined treatment of the fundamental problem as posed in our earlier paper.[1] Two new indices have been introduced as the means by which we have been able to bypass the Legendre transform of classical thermodynamics.

REFERENCES

1. Y. K. Huang, *Am. J. Phys.* **40**, 1261 (1972).
2. L. D. Landau and E. M. Lifshitz, *Statistical Physics* (Pergamon Press, London, 1958), pp. 165, 170, 174–175, and 184.
3. L. Knopoff, in *High Pressure Physics and Chemistry*, R. S. Bradley, ed. (Academic Press, New York, 1963), Vol. 1, pp. 227–245.
4. I. P. Bazarov, *Thermodynamics* (Pergamon Press, London, 1964), pp. 127–131.
5. Y. K. Huang, *J. Chem. Phys.* **53**, 571 (1970).
6. V. P. Kopyshev, *Soviet Phys.—Doklady* **10**, 338 (1965).

Reprinted from THE JOURNAL OF CHEMICAL PHYSICS, Vol. 45, No. 6, 1979–1984, 15 September 1966
Printed in U. S. A.

Thermodynamics of Shock Compression of Metals

Y. K. HUANG
Research Laboratory, Watervliet Arsenal, Watervliet, New York

(Received 18 April 1966)

This paper presents an analytic method for evaluating the effect of shock compression on metals. A lattice potential of the Born–Mayer type is used for calculating the pressure, compressibility, Grüneisen coefficient, entropy, temperature, and energy. A general equation of state is determined in the form of $p(v, T)=f(v)+Tg(v)$. Except for the use of a few well-established thermodynamic properties of the metal, no experimental data from shock-wave measurements are required for the theoretical calculation in this paper. Comparison between the calculated and cited experimental results is made only to test the accuracy of the analysis, and agreement turns out to be fairly close. The analysis also establishes a validity range for shock compressibility of solids, and the material model considered is sensitive to its lattice structure as compared with a continuum model.

INTRODUCTION

IT is of theoretical interest to consider the thermodynamics which underlies the shock compression of metals. When a metal is subjected to strong loading as imposed by explosion or high-speed impact, large plastic deformations are induced and the straining action turns out to be isotropic. Responses as such are due to the high rate and intensity of compressive stresses. Under strong shock compression the metal behaves at first plastically, but it is less sensitive to the strain rate. The material rigidity in shear may be neglected completely, and it is the compressibility that is responsible for the shock deformation. It should be noted that the metal may still possess its regular lattice structure even if it is under shock compression at several million atmospheres. The effect of shock propagation on the crystal lattice is to induce dislocations which would appear as some kind of misfit in a mosaic work. As a whole the crystal is still in a steady state of lattice vibration. All these considerations fix the concept of a material model which is different from the compressible fluidlike model for the so-called shock hydrodynamics of metals. The considered model of the metal is especially suitable for analysis to be incorporated with the Grüneisen–Debye theory of crystal-lattice dynamics.

Since the interatomic binding forces are essentially responsible for the metal behavior at low temperatures, it is expedient to consider the lattice potential and its several derived properties. In what follows, a potential function of the Born–Mayer type is used for such purposes. This form of lattice potential was first suggested by B. I. Davydov to be used for metals under shock compression with no phase transitions.[1] It is shown that the three constant coefficients of the Born–Mayer potential are otherwise determined in this paper by the use of a few well-established thermodynamic properties of the metal. By proper correlations the Born–Mayer potential function suffices to determine the pressure, compressibility, Grüneisen coefficient, entropy, temperature, and energy. A complete thermodynamic analysis can thus be carried out explicitly.

Considerations of rigidity and stability of the material turn out to impose limitation upon the shock compressibility of the metal. This is determined as the lower and upper bounds of the Grüneisen–Debye solid. Within the lower limit the theory of continuum mechanics is available for the description of the elastic- and plastic-wave behavior of the metal. Beyond the upper limit, a new model must be sought in order to account for the phase transition and quantum-electronic behavior of the metal. In this connection there exists a lack of both experimental and theoretical information concerning the shock compression of metals from 10 to 100 Mb (1 Mb$\approx 10^6$ atm) and from 10^3 to 10^4 °K.

LATTICE POTENTIAL AND RELATED PROPERTIES

A metal as a pure substance in equilibrium may be looked upon as a thermodynamic system which satisfies the basic relation

$$TdS=dE+pdv,$$

where T denotes the absolute temperature, S the specific entropy, E the specific energy, p the total pressure, and v the specific volume. Let $T=0$ or $dS=0$; then

$$p_c=-dE_c/dv.$$

This p_c is the so-called elastic pressure of the metal. The above equation determines the absolute-zero isotherm; it may also be looked upon as the isentropic equation of state. The elastic pressure will contribute almost completely to the total pressure for moderate compression, while the energy of the metal is composed of at least two parts: elastic and thermal. Only when the thermal energy E_D is increased to several times[2] as large as the elastic energy E_c, will the elastic pressure p_c deviate drastically from the total pressure p, and the difference between the total and elastic pressures

[1] L. V. Al'tshuler, Soviet Phys.—Usp. **8**, 52 (1965) [Usp. Fiz. Nauk **85**, 197 (1965)].

[2] Ya. B. Zeldovich and A. S. Kompaneets, *Theory of Detonation* (Academic Press Inc., New York, 1960), pp. 224, 228.

will, according to Grüneisen–Debye theory,[1] be called the thermal pressure p_D:

$$p_D = p - p_c = \gamma(E - E_c)/v = \gamma E_D/v$$
$$= 3R(\gamma/v)[TD(\theta/T) - T_0 D(\theta/T_0)],$$

with $\gamma = v(\partial p/\partial E)_v = -d(\ln\nu)/d(\ln v)$ = Grüneisen coefficient, $R = Nk/m$ = "gas" constant of metal, $\theta = h\nu_m/k$ = Debye temperature, $T_0 = 293°\text{K}$,

$$D\left(\frac{\theta}{T}\right) = 3\left(\frac{T}{\theta}\right)^3 \int_0^{\theta/T} \frac{y^3 dy}{e^y - 1} = \text{Debye function},$$

ν is the frequency of lattice vibration, N is Avogadro's number, k is Boltzmann constant, m is the atomic weight of metal, h is Planck's constant.

From the above discussion it is seen that an accurate determination of the elastic pressure is essential in what follows, yet the elastic pressure depends upon the lattice potential only. Let $x = v/v_0$, with $v_0 = 1/\rho_0$, the specific volume at zero pressure. Also, let A, B, and C denote the pertinent material constants to be determined in the sequel. The Born–Mayer lattice potential may be expressed as

$$E_c = A \exp(-Bx^{1/3}) - Cx^{-1/3}. \tag{1}$$

The elastic pressure is then

$$p_c = -dE_c/dv = \tfrac{1}{3}\rho_0 C x^{-2/3}\{\exp[B(1-x^{1/3})] - x^{-1/3}\}, \tag{2}$$

in which the equilibrium condition that $p_c = 0$ at $x = 1$ has been used in order to get $AB = C\exp(B)$. The bulk modulus as a measure of compressibility is given by

$$K = -v(dp_c/dv) = \tfrac{1}{9}\rho_0 C x^{-1/3}\{(B + 2x^{-1/3})$$
$$\times \exp[B(1-x^{1/3})] - 4x^{-1}\}, \tag{3}$$

$$K_0 = \tfrac{1}{9}\rho_0 C(B-2). \tag{3'}$$

Since it implies a cubic symmetry[3] of the lattice structure, the Dugdale–MacDonald formula is used in this paper for calculating the Grüneisen coefficient of the metal. Thus,

$$\gamma = -\frac{d(\ln\nu)}{d(\ln v)} = -\frac{1}{3} - \frac{v}{2}\frac{d^2(p_c v^{2/3})/dv^2}{d(p_c v^{2/3})/dv}$$

$$= \frac{1}{6}\left\{\frac{B^2 x^{1/3}\exp[B(1-x^{1/3})] - 6x^{-1}}{B\exp[B(1-x^{1/3})] - 2x^{-1}}\right\}, \tag{4}$$

$$\gamma_0 = \tfrac{1}{6}[(B^2-6)/(B-2)]. \tag{4'}$$

Equation (4) gives $\gamma = \gamma_0$ at $x = 1$, and $\gamma \to \frac{1}{2}$ as $x \to 0$. The latter limiting value of γ coincides with the quantum-statistical value of γ for an extremely compressed assembly of electrons.[1] Also, Eq. (4) offers a direct and quick determination of the Grüneisen coefficient at high pressures, while the corresponding calculation of γ based on shock-wave measurements is not so explicit.

From Eqs. (4′), (3′), and (2) follow

$$B = 3\gamma_0\{1 + [1 - (4\gamma_0 - 2)/3\gamma_0^2]^{1/2}\},$$

$$C = 9K_0/\rho_0(B-2),$$

and

$$A = C\exp(B)/B.$$

Thus, the material constants A, B, and C of Eq. (1) are completely determined from known values of the Grüneisen constant γ_0, density ρ_0, and bulk modulus K_0 at zero pressure. By the zero-pressure values are meant those values of properties which are measured at the normal atmospheric pressure, because 1 atm is negligible in comparison with the shock pressure. It should be noted that $B > 2$ because of Eq. (3′). So far, no dynamic property data are needed for our theoretical calculation, and such a calculation serves as a basis later to obtain the quasiequilibrium values of the shock-wave properties of the metal.

TEMPERATURE FUNCTION

It is of considerable interest to evaluate the shock temperature of the metal. In the literature, all calculated shock temperatures are based on the assumption that the entropy change in a solid may be neglected, but shock heating due to large compression has considerable effect on the lattice vibration and atomic packing. In this connection, two variable quantities have to be considered: the entropy rise across the shock front and the Grüneisen coefficient at high pressures. The entropy change is given[4] by

$$S - S_0 = (d^2 p_c/dv^2)_0 (v_0 - v)^3/12T_0$$
$$= C(B^2 + 6B - 18)(1-x)^3/324T_0. \tag{5}$$

Consider the entropy function $S = S(v, T)$ so that

$$dS = \left(\frac{\partial S}{\partial T}\right)_v dT + \left(\frac{\partial S}{\partial v}\right)_T dv = \frac{c_v dT}{T} + \left(\frac{\partial p}{\partial T}\right)_v dv,$$

$$= c_v d(\ln T) + (\partial p/\partial E)_v (\partial E/\partial T)_v dv,$$

$$= c_v[d(\ln T) + \gamma d(\ln v)],$$

$$= c_v d(\ln T) - c_v d\{\ln v^{1/3}[d(-p_c v^{2/3})/dv]^{1/2}\},$$

in which the relation[3]

$$\nu \sim v^{1/3}[d(-p_c v^{2/3})/dv]^{1/2}$$

has been used on account of cubic symmetry of the lattice. Here $c_v (\approx 3R)$ is the Dulong–Petit specific heat and ν is the frequency of lattice vibration. Integrating

---

[3] M. H. Rice, R. G. McQueen, and J. M. Walsh, Solid State Phys. **6**, 44–45 (1958).

[4] K. P. Stanyukovich, *Unsteady Motion of Continuous Media* (Pergamon Press, Ltd., London, 1960), p. 209.

and simplifying gives

$$T = T_0 x^{\frac{1}{3}} \left\{ \frac{B \exp[B(1-x^{\frac{1}{3}})] - 2x^{-1}}{B-2} \right\}^{\frac{1}{2}} \times \exp\left[\frac{C(B^2+6B-18)(1-x)^3}{324 c_v T_0} \right], \quad (6)$$

where T_0 may be taken as 293°K.

In closed form, Eq. (6) appears to depend upon the Born–Mayer lattice potential only. This is equivalent to saying that the shock temperature may be evaluated from its effect on the lattice vibration and atomic packing. It may be noted that Eq. (6) should be used only for the solid phase of the metal from T_0 to the melting point. Beyond this point both phase transition and electron specific heat will call for drastic correction of Eq. (6), because c_v and γ can no longer be treated in the same way as before.

It is interesting to determine the melting point of the metal under shock compression. According to the Lindemann law of fusion, the melting temperature of the metal under compression is given[5] by

$$T_m = x_m T_{m0} K_m / K_0, \quad (7)$$

where x_m is the root of the following algebraic equation:

$$(B-2)^{\frac{1}{2}} \{B \exp[B(1-x_m^{\frac{1}{3}})] - 2x_m^{-1}\}^{\frac{1}{2}} \times \exp[C(B^2+6B-18)(1-x_m)^3 / 324 c_v T_0] = (T_{m0}/T_0) x_m^{\frac{2}{3}} \{(B+2x_m^{-\frac{1}{3}}) \exp[B(1-x_m^{\frac{1}{3}})] - 4x_m^{-1}\}. \quad (8)$$

Table I. Basic data.

Quantity	Unit	Ag	Cu	Pb
ρ_0	g cm^{-3}	10.49	8.92	11.35
m	g	107.88	63.57	207.20
R	10^{-6} Mb g^{-1} cm^3(°K)$^{-1}$	0.77	1.31	0.40
θ	°K	229	343	96
T_{m0}	°K	1234	1356	600
γ_0		2.40	1.96	2.73
K_0	Mb	0.99	1.33	0.44
c_v	10^{-6} Mb g^{-1} cm^3 (°K)$^{-1}$	2.24	3.73	1.20
A	10^3 Mb g^{-1} cm^3	1.88	0.37	4.05
B		12.60	10	14.50
C	10^{-1} Mb g^{-1} cm^3	0.80	1.68	0.28
a	10^5 cm sec^{-1}	3.24	3.96	2.03
b		1.59	1.50	1.52
σ		0.37	0.33	0.45
c_L	10^5 cm sec^{-1}	3.60	4.75	2.10

[5] Yu. N. Ryabinin, K. P. Rodionov, and E. S. Alekseev, Soviet Phys.—Tech. Phys. **9**, 1477 (1965) [Zh. Techn. Fiz. **34**, 1913 (1964)].

Equation (8) is obtained by putting $T = T_m$ from Eqs. (6) and (7). Here, the subscript m designates melting; the subscript 0 denotes zero pressure; and T_{m0} is the melting point at atmospheric pressure. The density and pressure at melting can be determined readily. The complete determination of the Lindemann state thus marks off a phase transition of the metal at high pressure, and the Grüneisen–Debye theory is valid up to this point. It may be noted that the melting temperature and pressure of metals under shock compression are of the order of 10^3 °K and 1 Mb, respectively. The Fermi temperature is of the order of 10^4 °K for conduction electrons in the metal. It is known that above 10^4 °K and 100 Mb the Thomas–Fermi–Dirac theory is adequate to describe the predominant electronic behavior of the metal.[6] At present, the question is left open: how should the metal behavior be described in the range of 10^3–10^4 °K and 10–100 Mb?

VALIDITY RANGE FOR SHOCK COMPRESSIBILITY

As the Lindemann state imposes an upper limit upon shock compression of a solid, so will the material rigidity set a lower limit for the stability of the shock wave. Consideration of elastic and plastic waves in the solid leads to the criterion that a shock wave can remain stable only when its velocity is faster than the precursor elastic wave (whose amplitude is equal to the so-called Hugoniot elastic limit). Since the longitudinal elastic-wave velocity is given[7] by

$$c_L = [3K_0(1-\sigma)/\rho_0(1+\sigma)]^{\frac{1}{2}},$$

with σ equal to the Poisson ratio and, since the shock-wave velocity is shown later (a, b being constants) to be

$$D = a(1-b+bx)^{-1},$$

the lower limit of shock compressibility can be evaluated as

$$x_r = 1 - (c_L - a)/bc_L. \quad (9)$$

Here, the subscript r designates rigidity. Within this limit the stress-wave theory is available to account for the elastic and plastic behavior of the solid.

From the above discussion it follows that a shock-wave theory is valid for solids without involving phase transitions in the following range of compression:

$$x_r < x < x_m.$$

Beyond this validity range, different descriptions should be consulted from other material models.

[6] L. Knopoff, *High Pressure Physics and Chemistry* (Academic Press Inc., New York, 1963), Vol. 1, pp. 256–259.

[7] S. Timoshenko and J. N. Goodier, *Theory of Elasticity* (McGraw-Hill Book Co., New York, 1951), 2nd ed., pp. 10, 454.

TABLE II. Results for Ag.

A. Calculated[a]

x	p_c	γ	$S-S_0$	T	p_D	p	$E-E_0$
1.000	0	2.40	0	293	0	0	0
0.817	0.38	2.08	1.13(10^{-6})	725	0.026	0.41	3.80(10^{-3})
0.759	0.64	2.01	2.58(10^{-6})	1560	0.081	0.72	8.20(10^{-3})
0.710	0.97	1.95	4.50(10^{-6})	5150	0.256	1.23	18.3(10^{-3})
0.684	1.18	1.92	5.82(10^{-6})	7850	0.520	1.70	28.1(10^{-3})

B. Experimental

x	γ	$S-S_0$	T	p	$E-E_0$
1.000	2.65	0	293	0	0
0.817	1.76	1.17(10^{-6})	783	0.40	3.49(10^{-3})
0.759	1.84	2.66(10^{-6})	1621	0.70	8.05(10^{-3})
0.710	1.81	4.65(10^{-6})	2955	1.10	15.2(10^{-3})
0.684	1.77	6.00(10^{-6})	3998	1.40	21.1(10^{-3})

[a] Rigidity limit: $x_r=0.937$, $\rho_r=11.20$ g cm^{-3}, $\gamma_r=2.27$, $T_r=414$°K, $p_r=0.086$ Mb. Melting limit: $x_m=0.706$, $\rho_m=14.85$ g cm^{-3}, $\gamma_m=1.95$, $T_m=4420$°K, $p_m=1.15$ Mb.

GENERAL EQUATION OF STATE

From an earlier consideration, the general equation of state for the metal may be expressed as

$$p=f(v)+Tg(v)$$

with

$$f(v)=p_c-3p\gamma T_0RD(\theta/T_0),$$

$$g(v)=3\rho_0\gamma Rx^{-1}D(\theta/T). \qquad (10)$$

It should be noted that $D(\theta/T)$ is actually volume dependent, because from Eq. (6) T is a function of volume v only. Thus, Eq. (10) provides a complete thermodynamic description of the metal behavior under equilibrium condition. By use of suitable thermodynamic relations, other physical properties of the metal can be evaluated from Eq. (10).

The energy increase across the shock front is given by

$$E-E_0=E_c(x)+3RTD(\theta/T)$$
$$-[E_c(1)+3R_0T_0D(\theta/T_0)], \qquad (11)$$

where E_c is already given by Eq. (1).

CALCULATION AND RESULTS

Three metals, namely Ag, Cu, and Pb are chosen for comparative study. Basic data of Table I are calculated and taken from Grüneisen[8], Zemansky[9], and McQueen and Marsh.[10] Calculated results are listed in Tables II.A, III.A, and IV.A. Tables II.B, III.B and IV.B give the experimental results[10] corresponding to the calculated ones.

Experimental data[10] can also be reproduced by the following equations:

$$D=a+bu=a(1-b+bx)^{-1},$$

$$p=\rho_0a^2(1-x)(1-b+bx)^{-2},$$

$$E-E_0=(p/2\rho_0)(1-x),$$

$$S-S_0\approx(ba^2/3T_0)(1-x)^3,$$

$$T=T_0\exp\left(-\int_{v_0}^{v}\frac{\gamma dv}{v}\right), \qquad dS=0.$$

Here, D is the shock-wave velocity, u is the particle velocity, ρ_0 is the normal density; a and b are constants which are accurately determined from shock-wave measurements. It may be of interest to note that Berger and Joigneau[11] are able to give a rigorous proof of the linear law $D=a+bu$. They also show[11] that the theoretical values of a and b are given by

$$a=[v_0K_0(1+\alpha_0\gamma_0T_0)]^{\frac{1}{2}},$$

$$b=\tfrac{1}{2}(1+\alpha_0\gamma_0T_0)^{-1}\{\gamma_0+\tfrac{2}{3}+\alpha_0\gamma_0T_0[2\gamma_0-4-(7/3\gamma_0)]\},$$

where $v_0=1/\rho_0$, K_0 is the bulk modulus, γ_0 is the Grüneisen constant at $p=0$, and α_0 is the thermal coefficient of volume expansion at $T_0=293$°K. Calculated values of a and b turn out to agree closely with the corresponding experimental values.

From Tables II.A, II.B, III.A, III.B, IV.A, and IV.B it is seen that the calculated results agree fairly

[8] E. Grüneisen, "The State of A Solid Body," NASA Republ. RE-2-18-59W, 66 (1959).

[9] M. W. Zemansky, *Heat and Thermodynamics* (McGraw-Hill Book Co., New York, 1957), 4th ed., pp. 271, 274.

[10] R. G. McQueen and S. P. Marsh, J. Appl. Phys. **31**, 1253 (1960).

[11] J. Berger and S. Joigneau, Compt. Rend. **249**, 2506 (1959)

TABLE III. Results for Cu.

A. Calculated[a]

x	p_c	γ	$S-S_0$	T	p_D	p	$E-E_0$
1.000	0	1.96	0	293	0	0	0
0.814	0.46	1.72	$1.62(10^{-6})$	615	0.023	0.48	$5.46(10^{-3})$
0.762	0.71	1.65	$3.39(10^{-6})$	1085	0.059	0.77	$11.5(10^{-3})$
0.716	1.02	1.60	$5.78(10^{-6}$	2220	0.150	1.17	$18.8(10^{-3})$
0.690	1.24	1.57	$7.55(10^{-6})$	3640	0.266	1.51	$28.3(10^{-3})$

B. Experimental

x	γ	$S-S_0$	T	p	$E-E_0$
1.000	2.00	0	293	0	0
0.814	1.74	$1.71(10^{-6})$	719	0.50	$5.21(10^{-3})$
0.762	1.68	$3.59(10^{-6})$	1227	0.80	$10.7(10^{-3})$
0.716	1.62	$6.11(10^{-6})$	2301	1.20	$19.1(10^{-3})$
0.690	1.59	$7.98(10^{-6})$	3042	1.50	$26.0(10^{-3})$

[a] Rigidity limit: $x_r=0.889$, $\rho_r=10.03$ g cm^{-3}, $\gamma_r=1.78$, $T_r=390$°K, $p_r=0.224$ Mb. Melting limit: $x_m=0.687$, $\rho_m=12.98$ g cm^{-3}, $\gamma_m=1.56$, $T_m=4100$°K, $p_m=1.55$ Mb.

close with the experimental data as long as the shock compression lies within the melting limit. Beyond the Lindemann limit a large descrepancy between calculated and experimental results may be due to a phase transition and the quantum effect of electronic specific heat, both of which need further investigation.

CONCLUSION

As the Hugoniot relations are essential in any quantitative analysis of shock waves, so is the Grüneisen–Debye theory for the thermodynamic calculation of shock-wave properties of metals on the basis of quasi-equilibrium states. Both formulations are complementary to each other. This paper takes up the second approach in order to evaluate the effect of shock compression on metals.

A lattice potential of the Born–Mayer type is used in this paper for calculating the pressure, compressibility, Grüneisen coefficient, entropy, temperature, and energy. A general equation of state is also presented in the form of $p=f(v)+Tg(v)$. Other equations used in this paper are all derived from the lattice potential function and in explicit form. Except for the use of a few well-established thermodynamic properties of the

TABLE IV. Results for Pb.

A. Calculated[a]

x	p_c	γ	$S-S_0$	T	p_D	p	$E-E_0$
1.000	0	2.72	0	293	0	0	0
0.796	0.22	2.34	$0.69(10^{-6})$	873	0.023	0.24	$2.20(10^{-3})$
0.718	0.44	2.23	$1.85(10^{-6})$	2760	0.104	0.54	$6.40(10^{-3})$
0.673	0.66	2.16	$2.88(10^{-6})$	7370	0.310	0.97	$14.3(10^{-3})$

B. Experimental

x	γ	$S-S_0$	T	p	$E-E_0$
1.000	2.77	0	293	0	0
0.796	2.20	$0.61(10^{-6})$	901	0.20	$1.80(10^{-3})$
0.718	1.99	$1.61(10^{-6})$	1862	0.40	$4.97(10^{-3})$
0.673	1.86	$2.50(10^{-6})$	3739	0.60	$8.65(10^{-3})$

[a] Rigidity limit: $x_r=0.978$, $\rho_r=11.60$ g cm^{-3}, $\gamma_r=2.64$, $T_r=314$°K, $p_r=0.011$ Mb. Melting limit: $x_m=0.730$, $\rho_m=15.55$ g cm^{-3}, $\gamma_m=2.25$, $T_m=2300$°K, $p_m=0.36$ Mb

metal, no experimental data from shock-wave measurements are needed for the theoretical calculation in this paper. Comparison between the calculated and cited experimental results is made only to test the accuracy of the analysis, and agreement turns out to be fairly close.

Consideration of material rigidity and shock stability leads to a threshold compression below which the elastic- and plastic-wave theory is more adequate to describe the mechanical behavior of the metal. Shock heating causes melting of the metal at high pressures. This is determined as the Lindemann state which imposes an upper limit upon the Grüneisen–Debye theory for the metal. Beyond the Lindemann melting limit a new model must be sought in order to assess the phase transition and the electronic component of the shock-compression effect at high temperatures. In this connection there exists a lack of both experimental and theoretical information concerning shock compression of metals from 10–100 Mb and from 10^3–10^4 °K.

Reprinted from JOURNAL OF THE FRANKLIN INSTITUTE, Vol. 276, No. 1, July, 1963
Printed in U. S. A.

SHOCK DYNAMICS OF HYPERVELOCITY IMPACT OF METALS

BY

Y. K. HUANG[1] AND NORMAN DAVIDS[2]

ABSTRACT

Hypervelocity impact of metals poses a problem which has not yet received an analytic treatment. This paper presents as such a shock-dynamic analysis of the problem. Equations are developed in explicit form, to facilitate the determination of the shock waves produced during the impact process. Interaction, propagation, and stability of the shock waves are also discussed. The accuracy of the theory is finally tested by a comparative study of numerical examples and known experimental data.

1. STATEMENT OF THE PROBLEM

Hypervelocity in this paper means an impact velocity faster than the sound velocity in a material. Thus the sound velocity in metals is, in general, of the order of 10^5 cm/sec. When metals collide at speeds of this or higher order of magnitude, peculiar phenomena take place. In ballistic impact of a projectile against a target, strong loading waves are generated in both bodies, followed by the explosion-like yielding of the projectile and of partial target material. The latter effect creates an expanding shell from which an advancing spherical shock-front is detached. In the meantime, strong unloading waves are sent out from behind this shell and subsequently overtake the shock front, leaving permanent deformations over the crater surface on the one hand and causing shock decay on the other. An analytic treatment of this crater-formation problem has been given by the present writers (1),[3] on the basis of shock propagation in solids.

If one metal body collides with another over a relatively flat area of contact, strong compression and rarefaction shock-waves are produced at the interface. Interaction of these waves leads to the propagation of unsteady shock waves. The present paper analyzes shock formation, interaction and propagation during the impact process under consideration. The accuracy of the analysis will be tested by a comparative study of numerical examples and known experimental data.

2. SHOCK-WAVE BEHAVIOR OF SOLIDS

When a solid is subjected to strong loading such as a detonation or ballistic attack, deformation waves of large amplitudes will develop

[1] Associate Professor of Mechanical Engineering, South Dakota State College, Brookings, S. D.

[2] Professor of Engineering Mechanics, The Pennsylvania State University, University Park, Pa.

[3] The boldface numbers in parentheses refer to the references appended in this paper.

into shock waves because of the appreciable compressibility of the solid at high pressures. Under these extreme conditions of compression the stresses imposed on the material are exceedingly greater than its ultimate strength and the solid will, after undergoing plastic yielding, behave like a compressible fluid having negligible shearing resistance. Since it is a well-known theorem of hydrostatics that the stress tensor must be isotropic for an ideal fluid offering no resistance to shear, an analogous situation can be postulated here for the case of solids subjected to impact loading. As a result of this, the motion-stress equation of the solid based on nonlinear elasticity theory (2) takes the same form as the Euler equation of gas dynamics. Moreover, a nonlinear wave equation can be established which predicts the existence of shock waves in solids when their rigidity effect can be neglected and when the compressibility plays a role that governs the problem.

A second model for the shock-wave behavior of solids is as follows. In impact and similar transient processes, pressure waves propagate at speeds comparable to the sound velocity, and compressibility of the solid tends to amplify the pressure rise so rapidly that it results in a discontinuity of state across the wave front. Physically, waves of large amplitudes are affected by the impedance of the material in such a manner as to distort the waveform sharply until a shock front is formed to bridge two states of discontinuous properties.

When the states on both sides of a shock front are constant, the shock is said to be steady; otherwise, it is unsteady. Unsteady shock waves propagate with decreasing strengths and velocities. This phenomenon of shock decay presents one type of instability of shock waves. There are other types of unstable shock waves. Thus, a strong shock may induce b.c.c. → f.c.c. phase transition in metals such as iron, steel, bismuth, antimony, followed by breaking up into two or three shock waves (3). Plastic yielding of metals also causes multiple shock formation which is believed to be a result of the production of moving dislocations (4). Instability of shock waves can occur in metals whose Hugoniot curve is convex upward and/or includes cusps.

It should be pointed out that an analysis of shock stability is of fundamental importance in developing a theory which predicts the shock wave behavior of solids. For a given Hugoniot curve, the stability of a shock may be tested graphically by connecting the initial and final states of the shock with a straight line. If the line does not cut the Hugoniot curve at intermediate points, the shock is stable; otherwise, unstable. If the line lies above the Hugoniot curve, a compression shock bridges the two states; if it lies below, a rarefaction shock. When a shock is split into a series of shocks, the first will be the fastest and weakest, the second slower and stronger, and so on. In order to illustrate the above principles, it is of particular interest to examine the behavior of shock waves in iron or steel (5). Both iron and steel show a marked change in microstructure and a rapid increase in hardness when they are subjected to shock pressures above 130 kb. The multiple shock structure can be described as follows: (*a*) a single weak shock can exist only when its amplitude is smaller than 130 kb; (*b*) three shock waves, namely the elastic I, plastic I and plastic II, are generated from a strong shock whose amplitude lies between 130 and 330 kb.; (*c*) for pressure amplitudes between 330 and 664 kb., only the elastic I and plastic II waves are found; and (*d*) for shock pressures above 664 kb., only the plastic II wave is formed. A plastic II wave is any shock whose amplitude is larger than 130 kb.; the plastic I wave has an amplitude of 130 kb.; and the elastic I wave is defined by the Hugoniot elastic limit which is 6.8 kb. for iron and 10 kb. for steel.

3. SHOCK WAVES IN HYPERVELOCITY IMPACTS

When two metal bodies collide at hypervelocity and with considerable area in contact, strong shock waves that are nearly plane will develop rapidly in each body, due to compressibility effects on the propagation of large deformations at rather fast strain rates. Figure 1 shows two such bodies before collision, while Fig. 2 depicts the moment when they are in dynamic contact with the result that shock waves $\mathcal{S}_1$, $\mathcal{S}_2$ are formed adjacent to the interface. These waves cannot remain steady for any considerable duration, because unloading waves will soon be generated from the colliding surfaces where a sudden release of the impact pressure takes place and, propagating at faster speeds, will then overtake the shock waves, immediately causing decay of the latter. However, the shock waves $\mathcal{S}_1$, $\mathcal{S}_2$ can be considered as steady in the brief periods which extend from the instant of complete formation of the shock waves to the instant when they are interacted by the unloading waves. Thus, it is possible to make an analysis of $\mathcal{S}_1$, $\mathcal{S}_2$ as follows.

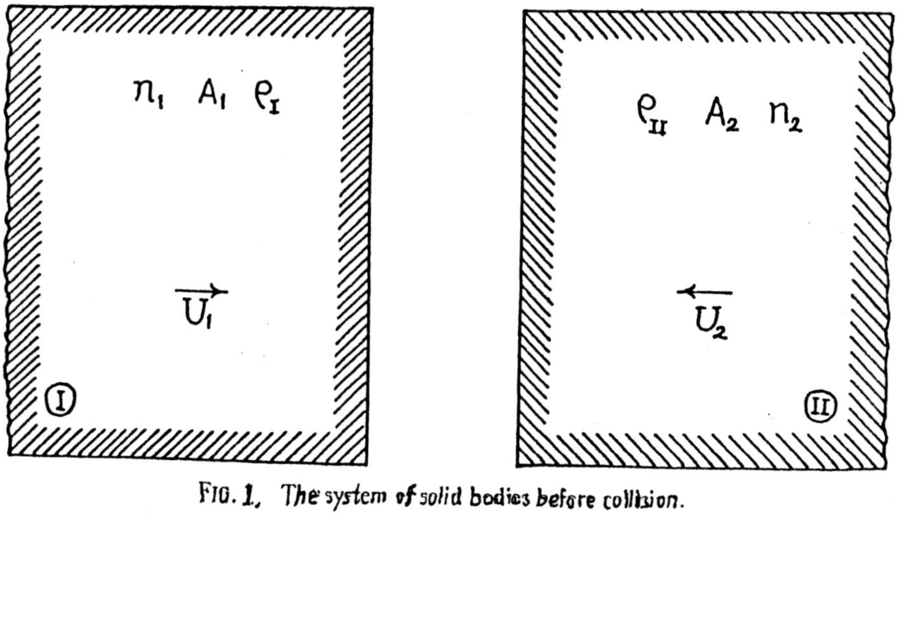

FIG. 1. The system of solid bodies before collision.

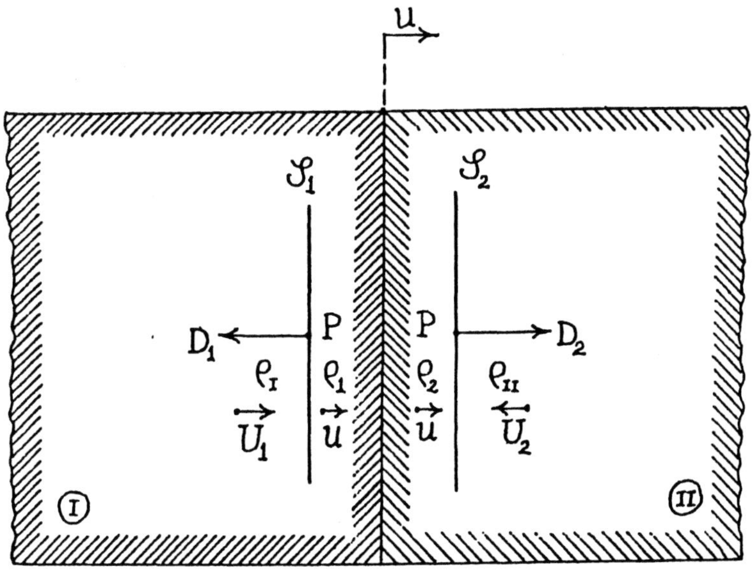

FIG. 2. Formation of shock waves upon impact.

The Hugoniot jump conditions can be written for S_1, S_2 giving six independent equations for the six unknowns (x, y, u, P, D_1, D_2):

$$\rho_1(D_1 + u) = \rho_I(D_1 + U_1) = j_1, \tag{a}$$

$$P + j_1(D_1 + u) = j_1(D_1 + U_1), \tag{b}$$

$$P = \frac{n_1 A_1 x}{1 - \frac{1}{2}(n_1 - 1)x}, \tag{c}$$

$$x = \frac{\rho_1}{\rho_I} - 1, \tag{d}$$

$$\rho_2(D_2 - u) = \rho_{II}(D_2 + U_2) = j_2, \tag{a'}$$

$$P + j_2(D_2 - u) = j_2(D_2 + U_2), \tag{b'}$$

$$P = \frac{n_2 A_2 y}{1 - \frac{1}{2}(n_2 - 1)y} \tag{c'}$$

$$y = \frac{\rho_2}{\rho_{II}} - 1. \tag{d'}$$

Equations (c) and (c') are the Hugoniot equations of state for shock compression as derived from the isentropic equation of state: $p = A\left\{\left(\frac{\rho}{\rho_0}\right)^n - 1\right\}$, for solids (6).

Algebraic manipulation of the above equations leads to:

$$\psi(y) = \frac{\rho_{II}(U_1 + U_2)^2}{2n_2A_2}\left\{\frac{2}{y^2} - \left(\frac{n_2 - 3}{y}\right) - (n_2 - 1)\right\}$$

$$= \left\{1 + \sqrt{\left(\frac{\rho_{II}}{\rho_I}\right)\frac{n_2A_2(1+y)}{n_1A_1 + \frac{1}{2}[n_2A_2(n_1+1) - n_1A_1(n_2-1)]y}}\right\}^2, \quad (1)$$

$$x = \frac{n_2A_2y}{n_1A_1 + \frac{1}{2}[n_2A_2(n_1-1) - n_1A_1(n_2-1)]y}, \quad (2)$$

$$u = \frac{U_1 - U_2\sqrt{\frac{\rho_{II}}{\rho_I}\left(\frac{x}{1+x}\right)\left(\frac{1+y}{y}\right)}}{1 + \sqrt{\frac{\rho_{II}}{\rho_I}\left(\frac{x}{1+x}\right)\left(\frac{1+y}{y}\right)}}, \quad (3)$$

$$P = \rho_{II}\left(\frac{1+y}{y}\right)\left\{\frac{U_1 + U_2}{1 + \sqrt{\frac{\rho_{II}}{\rho_I}\left(\frac{x}{1+x}\right)\left(\frac{1+y}{y}\right)}}\right\}^2, \quad (4)$$

$$D_1 = \frac{U_1 - (1+x)u}{x} = \sqrt{\frac{P}{\rho_I}\left(\frac{1+x}{x}\right)} - U_1, \quad (5)$$

$$D_2 = \frac{U_2 + (1+y)u}{y} = \sqrt{\frac{P}{\rho_{II}}\left(\frac{1+y}{y}\right)} - U_2. \quad (6)$$

If the two colliding bodies are made of the same material, then $n_1 = n_2 = n$, $A_1 = A_2 = A$, $\rho_I = \rho_{II} = \rho_0$, and Eq. 1 becomes

$$\left\{\frac{4nA}{\rho_0(U_1 + U_2)^2} + \frac{n-1}{2}\right\}y^2 + \left(\frac{n-3}{2}\right)y - 1 = 0.$$

Therefore

$$y = \frac{\sqrt{\left(\frac{n-3}{4}\right)^2 + \frac{4nA}{\rho_0(U_1+U_2)^2} + \frac{n-1}{2}} - \frac{n-3}{4}}{\frac{4nA}{\rho_0(U_1+U_2)^2} + \frac{n-1}{2}}, \quad (1a)$$

$$x = y, \quad (2a)$$

$$u = \tfrac{1}{2}(U_1 - U_2), \quad (3a)$$

$$P = \tfrac{1}{4}\rho_0\left(1 + \frac{1}{y}\right)(U_1 + U_2)^2, \tag{4a}$$

$$D_1 = \tfrac{1}{2}U_1\left(\frac{1}{y} - 1\right) + \tfrac{1}{2}U_2\left(\frac{1}{y} + 1\right), \tag{5a}$$

$$D_2 = \tfrac{1}{2}U_1\left(\frac{1}{y} + 1\right) + \tfrac{1}{2}U_2\left(\frac{1}{y} - 1\right). \tag{6a}$$

It is interesting to note two more special cases, namely, the case when $U_2 = 0$ and that when $U_1 = U_2$. Tests in which these are satisfied can readily be arranged by using impact machines and ballistic pendulums.

If the colliding velocities are not so high as to cause strong shock waves, the method of acoustic approximation (7) may be used to simplify the previous equations for further solution. Since $c^2 \approx \dfrac{\Delta p}{\Delta \rho}$ and $x, y < 1$, one may write $\dfrac{x}{1+x} \approx x \approx \dfrac{P}{\rho_I c_I^2}$, $\dfrac{y}{1+y} \approx y \approx \dfrac{P}{\rho_{II} c_{II}^2}$, where $c_I = \sqrt{\dfrac{n_1 A_1}{\rho_I}}$ and $c_{II} = \sqrt{\dfrac{n_2 A_2}{\rho_{II}}}$. Therefore,

$$y \approx \frac{\sqrt{\rho_I \rho_{II} \dfrac{n_1 A_1}{n_2 A_2}}}{\sqrt{n_1 A_1 \rho_I} + \sqrt{n_2 A_2 \rho_{II}}}(U_1 + U_2), \tag{1b}$$

$$x \approx \frac{\sqrt{\rho_I \rho_{II} \dfrac{n_2 A_2}{n_1 A_1}}}{\sqrt{n_1 A_1 \rho_I} + \sqrt{n_2 A_2 \rho_{II}}}(U_1 + U_2), \tag{2b}$$

$$u = \frac{U_1\sqrt{n_1 A_1 \rho_I} - U_2\sqrt{n_2 A_2 \rho_{II}}}{\sqrt{n_1 A_1 \rho_I} + \sqrt{n_2 A_2 \rho_{II}}}, \tag{3b}$$

$$P = \frac{\sqrt{n_1 A_1 n_2 A_2 \rho_I \rho_{II}}}{\sqrt{n_1 A_1 \rho_I} + \sqrt{n_2 A_2 \rho_{II}}}(U_1 + U_2), \tag{4b}$$

$$D_1 = \sqrt{\frac{n_1 A_1}{\rho_I}} - U_1, \tag{5b}$$

$$D_2 = \sqrt{\frac{n_2 A_2}{\rho_{II}}} - U_2. \tag{6b}$$

Equations 1*b* and 2*b* may be omitted, since (u, P, D_1, D_2) are now given explicitly in terms of material constants. For colliding bodies of the same material, further simplification of the above equations can readily be made. As an example, consider the determination of the impact pressure exerted by a plastic solid striking on a stationary, rigid wall. From Eq. 4*b*, it is clear that an "analytic" estimate gives $P = \rho_I c_I U_I$.

It should be noted that Eqs. 1 through 6 may also be applied to the analysis of shock reflection at an interface and of interaction between two shocks of different strengths.

4. NUMERICAL EXAMPLES

A survey of the literature indicates that little has been reported concerning the shock-wave behavior in hypervelocity impact of metal bodies that are initially stress-free. If the ballistic pendulum or impact machine is to be used for testing, the impact velocity must be boosted to the order of 10^6 cm/sec. This introduces difficulties in insuring structural strength for the equipment, although rocket power may be employed as an effective booster. Development of a shock-compression test by direct contact explosion leads to the technique of the momentum-transfer plate. This is a flying plate which is driven by a high explosive and strikes a metal target to produce a strong shock in the latter. It should be noted that the shock driver plate is actually not stress-free before collision with the target plate. However, a pressure-free state may be approximated by assuming that shock decay and pressure release are complete in the driver plate before the impact in question starts. Thus, in the following a few practicable sample problems will be considered in such a manner as to test the accuracy of the theory developed thus far.

Example 1

A shock-driver plate which is made of aluminum strikes a stationary, steel target-plate with a velocity of 1.63×10^5 cm/sec. Determine the pressure amplitudes and velocities of the shock waves produced by the hypervelocity impact, and compare the calculated results with corresponding experimental data.

As discussed earlier, the structure of a shock wave in steel varies with its pressure amplitude, and analysis of multiple shock formation must be made whenever iron or steel is concerned. This often requires revision of calculations as indicated below.

$$U_1 = 1.63 \times 10^5 \text{ cm/sec}, \; U_2 = 0,$$

$$n_1 = 3.55, \; A_1 = 232 \text{ kb}, \; \rho_I = 2.78 \text{ gm/cm}^3 \text{ (Al)},$$

$$n_2 = 3.53, \; A_2 = 480 \text{ kb}, \; \rho_{II} = 7.84 \text{ gm/cm}^3 \; (\alpha - \text{Fe}, P < 130 \text{ kb}),$$

$$n_2 = 3.80, \; A_2 = 339 \text{ kb}, \; \rho_{II} = 7.84 \text{ gm/cm}^3 \; (\gamma - \text{Fe}, P > 130 \text{ kb}).$$

From Eq. 1, it follows that

$$\psi(y) = \frac{0.123}{y^2} - \frac{0.033}{y} - 0.155 = \left(1 + \sqrt{\frac{1+y}{0.173 + 0.588y}}\right)^2.$$

Therefore,

$$y = 0.108, \qquad x = 0.194, \qquad P = 212 \text{ kb}.$$

Thus, a three-wave structure is associated with the shock in the steel target. The elastic I wave has a pressure amplitude of 10 kb and a velocity $c_o = \sqrt{\frac{E}{\rho_0}} = 5.13 \times 10^5$ cm/sec. The plastic I and II waves are determined as follows. Applying the Hugoniot jump conditions, Eqs. *a* and *b* modified, to the plastic I wave, results in

$$7.84D_{p1} = 8.38(D_{p1} - u'),$$
$$130 \times 10^9 = 7.84D_{p1}u',$$

Therefore,

$$D_{p1} = 5.07 \times 10^5 \text{ cm/sec}, \qquad u' = 0.328 \times 10^5 \text{ cm/sec}.$$

From Eqs. 1, 2, 4, and d',

$$\psi(y) = \frac{0.161}{y^2} - \frac{0.065}{y} - 0.226 = \left(1 + \sqrt{\frac{1+y}{0.266 + 4.88y}}\right)^2.$$

Therefore,

$$y = 0.130, \qquad x = 0.189, \qquad P = 205 \text{ kb.},$$
$$\rho_2 = (1 + y)\rho_{II} = 8.859 \text{ gm/cm}^3.$$

Applying the Hugoniot jump conditions, Eqs. *a* and *b* modified, to the plastic II wave gives

$$8.38(D_{p2} - 0.328 \times 10^5) = 8.859(D_{p2} - u) = j,$$
$$(205 - 130) \times 10^9 = j(u - 0.328 \times 10^5).$$

Therefore,

$$D_{p2} = 4.54 \times 10^5 \text{ cm/sec}, \qquad u = 0.54 \times 10^5 \text{ cm/sec}.$$

The single shock wave in the driver plate can be determined similarly by use of Eqs. 3, 4, and 5. The calculated results are compared with experimental data in Table I, indicating close agreement.

TABLE I.—*Comparison of Theoretical and Experimental Results for Al-Fe Impact.*

Shock Waves in Target	Theoretical	Experimental (8)
Elastic I Wave		
P''	10 kb	—
ρ''	7.87 gm/cm³	—
D_{e1}	5.13×10^5 cm/sec	—
Plastic I Wave		
P'	130 kb	130 kb
ρ'	8.38 gm/cm³	8.38 gm/cm³
D_{p1}	5.07×10^5 cm/sec	5.09×10^5 cm/sec
u'	0.328×10^5 cm/sec	—
Plastic II Wave		
P	205 kb	201 kb
ρ_2	8.86 gm/cm³	—
D_{p2}	4.54×10^5 cm/sec	3.77×10^5 cm/sec
u	0.54×10^5 cm/sec	0.57×10^5 cm/sec

Example 2

In the determination of shock-dynamic properties of iron, both the shock driver and target specimen are made of iron. Given the impact velocity $U_1 = 8.4 \times 10^5$ cm/sec ($U_2 = 0$), it is required to determine the shock pressure, shock velocity, particle velocity, and density behind the shock front.

Applying Eqs. 1*a*, 2*a* and 4*a* with $n = 3.53$, $A = 480$ kb, $\rho_0 = 7.84$ gm/cm³, gives $x = y = 0.582$, $P = 3740$ kb. According to Section 2, a single strong shock (that is, the plastic II) wave is produced in the target metal. Thus, a revision of the calculation is necessary with $n = 3.80$, $A = 339$ kb, $\rho_0 = 7.84$ gm/cm³. From Eqs. 1*a*, 4*a*, 6*a*, and 3*a*, respectively, are obtained

$$y = \frac{\sqrt{(0.2)^2 + 0.933 + 1.4} - 0.2}{0.933 + 1.4} = 0.575,$$

$$P = \tfrac{1}{4}(7.84)\left(\frac{1.575}{0.575}\right)(8.4 \times 10^5)^2 \div 10^9 = 3780 \text{ kb}.$$

$$D = \frac{1}{2}\left(\frac{1.575}{0.575}\right)(8.4 \times 10^5) = 11.50 \times 10^5 \text{ cm/sec},$$

$$u = \tfrac{1}{2}(8.4 \times 10^5) = 4.20 \times 10^5 \text{ cm/sec},$$

$$\rho = 1.575 \times 7.84 = 12.35 \text{ gm/cm}^3.$$

The above results are compared with experimental data in Table II, indicating close agreement.

TABLE II.—*Comparison of Theoretical and Experimental Results for Fe-Fe Impact.*

Strong Shock in Target	Theoretical	Experimental (9)
U (Impact Velocity)	8.4×10^5 cm/sec	8.4×10^5 cm/sec
P	3780 kb	3440 kb
D	11.50×10^5 cm/sec	10.45×10^5 cm/sec
u	4.20×10^5 cm/sec	4.20×10^5 cm/sec
ρ	12.35 gm/cm^3	13.13 gm/cm^3

Example 3

It is estimated that the impact velocity is 3.42×10^5 cm/sec when a flying aluminum plate strikes an aluminum target plate. Determine the shock-dynamic properties of aluminum from this impact process.

Applying Eqs. *1a*, *2a*, *4a*, *6a* and *3a* with $n = 3.55$, $A = 232$ kb, $\rho_0 = 2.78$ gm/cm^3, gives

$$x = y = 0.284,$$

$$P = 379 \text{ kb}, \qquad \rho = 3.57 \text{ gm/cm}^3,$$

$$D = 7.95 \times 10^5 \text{ cm/sec}, \qquad u = 1.71 \times 10^5 \text{ cm/sec},$$

$$c = \sqrt{\frac{n(A + P)}{\rho}} = 7.8 \times 10^5 \text{ cm/sec}.$$

Table III shows the above results in comparison with experimental data, indicating close agreement.

TABLE III.—*Comparison of Theoretical and Experimental Results for Al-Al Impact.*

Shock Wave in Target	Theoretical	Experimental [10]
U (Impact Velocity)	3.42×10^5 cm/sec	3.42×10^5 cm/sec
P	379 kb	346.2 kb
D	7.95×10^5 cm/sec	7.49×10^5 cm/sec
u	1.71×10^5 cm/sec	1.712×10^5 cm/sec
ρ	3.57 gm/cm^3	3.5 gm/cm^3
c	7.8×10^5 cm/sec	7.842×10^5 cm/sec

Example 4

A commercial lead bullet strikes the pressure end of a Hopkinson bar with a velocity of 3.65×10^4 cm/sec. Determine the peak impact pressure and compare it with the experimental result.

Apply Eq. 4*b* with $n_1 = 3.11$, $A_1 = 183$ kb, $\rho_I = 11.34$ gm/cm^3, $n_2 = 3.53$, $A_2 = 480$ kb, $\rho_{II} = 7.84$ gm/cm^3, $U_1 = 3.65 \times 10^4$ cm/sec, $U_2 = 0$.

$$P = \frac{(2.54 \times 10^6)(3.65 \times 10^6)}{2.54 \times 10^6 + 3.65 \times 10^6} \times (3.65 \times 10^4) \div 10^9 = 54.6 \text{ kb}.$$

The corresponding experimental result (11) is reported to be 15.4 kb. Here the large deviation may be due to the entirely different physical models on which the propagation of waves is based. But the calculated result gives only a first approximation to the headmost portion of the curved shock front produced by the projectile in the target.

The foregoing sample problems have been treated in detail with a view to applying the theory developed thus far. The calculated results are then compared with corresponding experimental results borrowed from the literature. Examples 1 through 3 indicate unusually close agreement of theoretical and experimental results. This serves to confirm the proposed theory for the plane shock-wave behavior in normal impact. It is seen that the closer the colliding bodies are to unbounded media, the more accurate results will be given by Eqs. 1 to 6. Thus, Example 4 leads to peculiar discrepancies because of inconsistent boundary conditions as well as the fact that a different theoretical basis applies. The consequence of this is twofold: first, a certain versatility is expected of the theory; second, a limitation is imposed on implications of the theory.

5. CONCLUDING REMARKS

The foregoing analysis predicts shock-wave behavior in aluminum-steel, steel-steel and lead-steel impacts giving values in close agreement with known experimental data. This should have a bearing on further developments of high-pressure physics. Thus, by the technique of metal-to-metal impact at hypervelocity, new experimental arrangements are possible which not only remove the difficulty of high-pressure packing but also yield much higher pressures, say, up to several megabars. Moreover, this investigation has engineering applications. For instance, new impact experiments can be carried out to obtain more information on the dynamic behavior of materials, and the new data thus obtained can serve as a guide to the manufacture of new engineering materials and to the design of equipment such as armor plate or structural members likely to be subjected to ballistic or explosive attack.

REFERENCES

(1) N. DAVIDS AND Y. K. HUANG, *J. Aero. Sci.*, Vol. 29, p. 550 (1962).
(2) V. V. NOVOZHILOV, "Theory of Elasticity," London, Pergamon Press, 1961, p. 108.
(3) W. E. DRUMMOND, *J. Appl. Phys.*, Vol. 28, p. 998 (1957).
(4) W. J. BAND, *Geophys. Research*, Vol. 65, p. 695 (1960).
(5) Metallurgical Society Conferences Vol. 9, "Response of Metals to High Velocity Deformation," New York, Interscience, 1961, pp. 249–274.
(6) Y. K. HUANG, "Shock Waves in Hypervelocity Impact of Metals," Ph.D. Thesis, The Pennsylvania State University, 1962, p. 14.
(7) K. P. STANYUKOVICH, "Unsteady Motion of Continuous Media," London, Pergamon Press, 1960, p. 571.
(8) J. O. ERKMAN, *J. Appl. Phys.*, Vol. 32, p. 939 (1961).
(9) L. V. ALTSHULER, K. K. KRUPNIKOV, B. N. LEDENEV, V. I. ZHUCHIKNIN, AND M. I. BRAZHNIK, *Soviet Phys. JETP*, Vol. 34, p. 606 (1958).
(10) H. D. MALLORY, *J. Appl. Phys.*, Vol. 26, p. 555 (1955).
(11) R. M. DAVIES, *Phil. Trans. Roy. Soc., London*, Vol. A-240, p. 375 (1948).

APPENDIX

Symbols

A, A_1, A_2	constant in the isentropic equation of state for metals, kb
c	sound velocity, cm/sec
c_0	elastic wave velocity, cm/sec
D, D_i	shock wave velocity ($i = 1, 2, e1, p1, p2$), cm/sec
E	Young's modulus of elasticity, kb
j, j_1, j_2	mass flux density, gm/cm^2 sec
n, n_1, n_2	adiabatic index in the isentropic equation of state for metals
p	thermodynamic pressure, kb
P	Hugoniot pressure, kb
S_1, S_2	shock waves
u	particle velocity, cm/sec
U_1, U_2	impact velocity, cm/sec
ρ, ρ_i	density (i = 0, 1, 2, I, II), gm/cm^3

Index

A

ablating plasma, 118
acoustic impedance, 6, 89
acoustic velocity, 6, 42
aerodynamics, 20
Alfvén wave, 101
analogy, 16
argon, 2, 110
armor, 116
artificial viscosity, 137
astrophysics, 45, 46
atomic bomb (A-bomb), 118
atomic number (Z), 78
Avogadro number, 77

B

ballistic shock, 2
ballistics, 111
    interior ballistics, 55
    penetration ballistics, 111
    terminal ballistics, 115
basic equation of gasdynamics, 136
basic shock jump conditions, 4
Bernoulli equation, 116
binodal (saturation curve), 87
BKW equation of state, 93
Bohr radius of hydrogen atom, 76, 78
Boltzmann constant, 44, 45
Boltzmann distribution, 43, 46
Boltzmann equation, 132
Boltzman law, 43
boson, 52
bottleneck (traffic flow), 130
Boyle's law, 124
bulk modulus
    isentropic bulk modulus, 50
    isothermal bulk modulus, 42
Burgers equation, 132
Bridgman isotherm, 57

C

cavitation (hydrodynamics), 127
celerity (water wave velocity), 81
Clasusius Virial theorem, 52
computational fluid dynamics (CFD), 136, 138, 139
computer code, 138
computer simulation, 135
condensed matter, 49
conductivity
    electric conductivity, 100
    thermal conductivity, 134
continuity, 27
conservation of mass, momentum, and energy, 4
coordinates, 4
    laboratory coordinate (at rest), 5
    moving coordinate (for shock flow), 5
co-volume (van der Waals), 54
crater, 112
cratering efficiency, 115
critical point/state, 54
    critical pressure, 54
    critical temperature, 54
    critical volume, 54
crystal
    covalent crystal, 64
    ionic crystal, 64
    metallic crystal, 64
    molecular crystal, 64
    crystalline solids, 64

D

damage number, 111
Debye solid (phonon), 52
Debye theory, 59
degenerate gas, 52
degenerate electron, 52
degree of freedom, 48
density wave (traffic flow), 128
detonation, 88
    Chapman-Jouguet (C-J) detonation, 89, 90

converging detonation wave, 89
detonation products (DP), 49, 88
detonation wave, 88
overdriven detonation, 89
strong detonation, 89
weak detonation, 89
deuterium, 1, 118
discontinuity, 4
contact discontinuity, 108
shock discontinuity/jump, 10
distensibility, 69
distention, 69
distribution (statistical), 43
Boltzmann distribution, 43
Bose-Einstein distribution, 43
canonical distribution, 43
Fermi-Dirac distribution, 43
Gibbs distribution, 43
grand canonical distribution, 43
microcanonical distribution, 43
downstream, 4

E

electric field, 100
electromagnetic theory, 99
electron
completely degenerate electron, 52
conduction electron, 53
free electron, 53
ellipsoidal reflector, 126
enthalpy, 4, 50, 68
entropy jump, 13
entropy production, 86, 132
equation of dispersion, 128
equation of state (EOS), 3, 4, 6, 39, 45, 50, 51, 67, 78, 131, 140
Able EOS, 52, 55, 92
BKW EOS, 93
Bridgeman EOS (isotherm), 57
classical ideal gas EOS, 45
Carnahan-Starling EOS, 92
γ-law gas EOS, 7, 49
Greenberg EOS, 131
Greenshield EOS, 128
Grüneisen EOS, 42, 59, 60, 94
Hugoniot EOS, 13, 97
isentropic EOS, 7
isothermal EOS, 82, 124
JWL EOS, 93, 94
Mie-Grüneisen EOS, 59, 60, 72
Murnaghan EOS (isentrope), 56, 66
Percus-Yevick EOS, 92
quantum ideal gas EOS, 51, 96
shock adiabat, 13
Tait EOS, 56, 57
Thomas-Fermi EOS, 4, 53, 74, 76, 77
Thomas-Fermi-Dirac EOS, 77
Van der Waals EOS, 53, 55
virial EOS, 93
EOS indices (g, j), 51
Eulerian code, 138
explosion
chemical explosion, 22
dust explosion, 88
explosion effect, 112
nuclear explosion, 22
point explosion, 23
vapor explosion, 87
extracorporeal shock wave lithotripsy (ESWL), 125

F

Fermat principle, 26
Fermi-Dirac distributions, 43, 75
Fermi energy, 52, 74
Fermi gas, 53
Fermi momentum, 78
fermion, 52
finite difference (FD)
FD grid/mesh, 136
FD method, 138
finite element method, (FEM), 140
fireball, 125
flow
hypersonic flow, 20
incompressible flow, 81
plasma flow, 99
shock flow, 135
subsonic flow, 20
supersonic flow, 20
transonic flow, 20
fluid dynamics, 99
FORTRAN format, 136
FORTRAN instruction, 138
Fraunhofer diffraction, 26
Froude number, 81
Fresnel diffraction, 26

G

gas
ideal gas, 7, 96
γ-law gas, 7
Navier-Stokes-Fourier gas, 134
gasdynamics, 1
gasdynamic shock wave, 3
gaskinetics (kinetic theory of gas), 48, 132, 134, 135
generalization, 16

Gibbs free enthalpy, 50, 68
Grüneisen parameter Γ, 9, 42, 68
Dugdale-MacDonald Γ, 61
Slater Γ, 61

Vashchenko-Zubarev Γ, 61
Greenshield model, 128, 129, 130

H

heat conduction, 134
helium, 2, 46
Helmholtz free energy, 44
high explosive (HE), 88
    high-explosive anti-tank (HEAT) round, 117
    high-explosive-plastic (HEP), 117, 118
Hugoniot curve (Rankine-Hugoniot curve), 70, 87, 88
    DP Hugoniot, 88
    HE Hugoniot, 88
hydraulic jump, 81
hydrogen ${}_1^1H_0$, 46, 94
    deuterium ${}_1^2H_1$, 118
    tritium ${}_1^3H_2$, 118
hypervelocity impact, 111, 112, 114

I

ideal gas (classical, quantum), 43, 45, 48-51
impact, 111
    ballistic impact, 111
    elastic impact, 111
    explosive impact, 111
    hypervelocity impact and catering, 115
    impact effect, 112
    liquid-to-solid impact, 112
    meteoric impact and catering, 140
    plastic impact, 111
    solid-to-solid impact, 112
inertial confinement fusion (ICF), 1-3, 118, 120, 135
interaction potential, 59
internal energy, 4
ionization, 47, 94
    complete/full ionization, 47
    degree of ionization, 47
    ionization front, 122
    ionization potential, 47
    partial ionization, 47

J

jet
    shaped-charge jet penetration, 116
    stand-off (shaped-charge), 116

K

kidney stone (ESWL), 125, 127, 135
kinematic shock (traffic flow), 3, 128
kinetic energy, 52
kink, 87

L

Lagrangian code, 138
Larmor radius, 104
lattice
    latticework, 41
    lattice potential, 64
laser beam, 104
laser-supported detonation (LSD), 3, 4, 104, 105
lawson criterion for nuclear fusion, 119
light-gas gun, 112
lithotripter, 125, 126
Lorentze force, 94, 100

M

Mach number (shock), 3, 13, 81, 97, 109, 125
Mach reflection, 20, 32-34, 57, 64, 111
magnetic field, 94, 100
magnetic induction, 100
magnetic permeability, 100
magnetic pressure, 100
magnetic tension, 100
magnetohydrodynamics (MHD), 2, 41, 99, 107
magnetosonic waves (fast, intermediate, slow), 101
mass number, 78
mean free path, 132
mirror-image approximation, 114
mixture (theory of mixture), 68
MKSA units (meter-kilogram-second-ampere), 100
molecular collision, 132
molecular dynamics (MD), 140
Monte Carlo simulation, 140
Murnaghan isentrope, 56

N

Neumann spike (von Neumann spike), 89
neutron number, 78
number density (n), 46
numerical modeling, 135

O

Ohm's law, 100

P

partition function, 43
penetration depth, 115
phonon, 52
photon, 52, 96
planck constant ($\hbar$), 46, 78
plastic
    Plexiglas (PMMA), 67
    teflon, 67
Poisson equation, 74, 75
Pointing vector, 100
polymer, 68
porosity, 69, 71
porous material, 67, 69
potential function, 75
powder, 67, 69, 73
Prandh-Meyer expansion, 34
pressure
    artificial viscous pressure, 137
    magnetic pressure, 100
    radiation pressure, 96
    shock pressure, 11
proton, 79

Q

quantized (degenerate), 50
quantum ideal gas, 50

R

Rankine-Hugoniot equation, 17, 72
Rankine-Hugoniot relation, 3, 5, 63, 128, 141
rarefied gas, 52
rarefaction wave, 10, 108
Rayleigh line, 88
reflection (shock reflection), 26
    irregular reflection (IR), 36
    Mach reflection (MR), 32-34
    oblique reflection, 36
    regular reflection (RR), 33, 36
refraction (transmission), 26
retrograde fluid (negative shock), 11, 86
Riemann invariant, 31, 109
road traffic, 127, 128
roton (quantized vortex), 53

S

Saha theory, 41
Saha equation, 47
shaped charge, 116
shock
    shock angle, 16
    shock compaction, 1, 72
    shock compression, 10, 119
    shock decay, 25
    shock deflection, 16, 40
    shock diffraction, 19, 26, 32, 38, 135
    shock EOS, 58, 64, 86
    shock flow, 26, 135
    shock focusing, 125
    shock front, 4
    shock heating, 10, 11, 63, 71, 132
    shock Hugoniot (linear quadratic), 15, 71, 91
    shock Hugoniots, 58, 59, 69
    shock impedance, 29, 88
    shock jump (discontinuity), 4, 10, 96, 134
Snell's laws, 26, 37
    shock layer (shock transition), 23
    shock path, 130
    shock phenomenon, 1, 2
    shock pressure, 11, 121
    shock properties, 10
    shock reflections, 13, 19, 26, 27, 29, 30, 34,40
    shock refraction/transmission, 26, 29
    shock stability, 10, 11, 40, 104
    shock strength, 31, 57, 125
    shock structure (shock profile), 132, 133
    shock temperature, 11, 95
    shock thickness, 124, 132
    shock trajectory, 24, 106
    shock travel time, 24
    shock tube, 2, 107-109
shock tunnel (hypersonic), 110
shock velocity (U), 3, 40
shock wave, shock, 1, 3, 9, 26, 120
    bow shock, 2, 32
    collisionless shock (MHD), 101, 104
    compression shock (positive shock), 3
    condensation shock, 87
    converging shock, 3
    cosmic shock, 121
    curved shock, 3
    cylindrical shock, 3, 28
    detached shock, 32, 36
    diverging shock, 3, 29
    evaporation shock, 87
    fast shock (MHD), 99, 104
    gravity shock, 82
    hydromagnetic shock (MHD), 99, 101, 104
    imploding shock, 2, 28
    inert shock, 3
    intermediate shock (MHD), 104
    ionizing shock, 94
    isothermal shock, 124
    laser shock, 104
    liquefaction shock, 87
    luminous shock, 104
    MHD shock, 104
    normal shock, 13, 22, 32, 38, 39
    oblique shock, 16, 17, 20, 21, 32, 33, 37
    parallel shock, (MHD), 102

perpendicular shock (MHD), 102, 103, 111
plane shock, 3
plasma shock, 2, 4
radiating shock, 96
rarefaction shock (negative shock), 3, 12, 82, 87
slow shock (MHD), 104
spherical shock, 3 23, 28
steady shock (uniform shock), 132
strong shock, 16
strong shock limit, 15, 48
switch-off shock (MHD), 104
switch-on shock (MHD), 104
underwater shock, 1, 81, 83
weak shock, 16, 32
self-similar flow, 22
slip stream (vortex sheet), 37
snow-plow model, 69, 70, 72
sonic boom, 2
sound speed (sonic velocity), 13
specific heat at constant pressure (*cp*), 6
specific heat at constant volume (*cv*), 6
specific heat ratio, 7
states of matter (gas, liquid, solid, plasma), 41
critical state (*pc*, *vc*, Tc), 54
shocked state (*ph*, *vh*, TH), 8
statistical mechanics, 6, 43
superfluid (helium at low temperature), 53
supernova, 121

T

Taylor series expansion, 8
temperature, 5
tensor, 100
thermal conductivity, 134
thermodyamics, 6, 41, 79
thermonuclear reaction, 119
Thomas-Fermi-Dirac theory, 77
Thomas-Fermi theory, 74, 79
traffic density, 127, 130
traffic flow, 127
traffic jam, 127, 130, 131
tritium ($^3_1H_2$), 1, 118

U

ultra-high pressure (> 10 Mb), 4
underground nuclear test, 4
upstream, 4

V

vapor explosion, 87
vector, 100
velocity
flow velocity (*u*), 4
group velocity, 128
phase velocity (wave velocity), 128
virial (Clausius virial), 52
virial EOS, 93
viscosity, 1, 83, 99, 134
vortex, (vorticity), 32
van der Waals fluids, 53, 87
von Neumann spike, 88, 89

W

warhead, 116
wave
compression wave, 10
density wave (traffic flow), 128
linear wave, 26
nonlinear wave, 26, 128
rarefaction wave, 10
shock wave, 1, 36
wavelength, 128
wave number, 128
weaponry, 111
Whitman's GSD, 38